Photoactive Materials

Photoactive Materials
Synthesis, Applications and Technology

Editor

Pierre-Alexandre Blanche

MDPI • Basel • Beijing • Wuhan • Barcelona • Belgrade • Manchester • Tokyo • Cluj • Tianjin

Editor
Pierre-Alexandre Blanche
Wyant College of Optical Sciences,
The University of Arizona
USA

Editorial Office
MDPI
St. Alban-Anlage 66
4052 Basel, Switzerland

This is a reprint of articles from the Special Issue published online in the open access journal *Materials* (ISSN 1996-1944) (available at: https://www.mdpi.com/journal/materials/special_issues/photoactive_materials).

For citation purposes, cite each article independently as indicated on the article page online and as indicated below:

LastName, A.A.; LastName, B.B.; LastName, C.C. Article Title. *Journal Name* **Year**, *Volume Number*, Page Range.

ISBN 978-3-0365-0958-7 (Hbk)
ISBN 978-3-0365-0959-4 (PDF)

© 2021 by the authors. Articles in this book are Open Access and distributed under the Creative Commons Attribution (CC BY) license, which allows users to download, copy and build upon published articles, as long as the author and publisher are properly credited, which ensures maximum dissemination and a wider impact of our publications.

The book as a whole is distributed by MDPI under the terms and conditions of the Creative Commons license CC BY-NC-ND.

Contents

About the Editor vii

Pierre-Alexandre Blanche
Special Issue: Photoactive Materials: Synthesis, Applications and Technology
Reprinted from: *Materials* 2021, 14, 585, doi:10.3390/ma14030585 1

Jae Yong Jung, Juna Kim, Yang Do Kim, Young-Kuk Kim, Hee-Ryoung Cha, Jung-Goo Lee, Chang Sik Son and Donghyun Hwang
Enhanced Crystallinity and Luminescence Characteristics of Hexagonal Boron Nitride Doped with Cerium Ions According to Tempering Temperatures
Reprinted from: *Materials* 2021, 14, 193, doi:10.3390/ma14010193 5

Pierre-Alexandre Blanche, Adoum H. Mahamat and Emmanuel Buoye
Thermal Properties of Bayfol® HX200 Photopolymer
Reprinted from: *Materials* 2020, 13, 5498, doi:10.3390/ma13235498 15

V. A. Kotov, M. Nur-E-Alam, M. Vasiliev, K. Alameh, D. E. Balabanov and V. I. Burkov
Enhanced Magneto-Optic Properties in Sputtered Bi- Containing Ferrite Garnet Thin Films Fabricated Using Oxygen Plasma Treatment and Metal Oxide Protective Layers
Reprinted from: *Materials* 2020, 13, 5113, doi:10.3390/ma13225113 27

Pannaree Srinoi, Maria D. Marquez, Tai-Chou Lee and T. Randall Lee
Hollow Gold-Silver Nanoshells Coated with Ultrathin SiO_2 Shells for Plasmon-Enhanced Photocatalytic Applications
Reprinted from: *Materials* 2020, 13, 4967, doi:10.3390/ma13214967 41

Danil A. Kurshanov, Pavel D. Khavlyuk, Mihail A. Baranov, Aliaksei Dubavik, Andrei V. Rybin, Anatoly V. Fedorov and Alexander V. Baranov
Magneto-Fluorescent Hybrid Sensor $CaCO_3$-Fe_3O_4-$AgInS_2$/ZnS for the Detection of Heavy Metal Ions in Aqueous Media
Reprinted from: *Materials* 2020, 13, 4373, doi:10.3390/ma13194373 53

Zhiguo Zhao and Xue Li
Hydrothermal Synthesis and Optical Properties of Magneto-Optical Na_3FeF_6:Tb^{3+} Octahedral Particles
Reprinted from: *Materials* 2020, 13, 320, doi:10.3390/ma13020320 65

Boris I. Lembrikov, David Ianetz and Yosef Ben-Ezra
Nonlinear Optical Phenomena in a Silicon-Smectic A Liquid Crystal (SALC) Waveguide
Reprinted from: *Materials* 2019, 12, 2086, doi:10.3390/ma12132086 73

Eric Abraham Hurtado-Aviles, Jesús Alejandro Torres, Martín Trejo-Valdez, Christopher René Torres-SanMiguel, Isaela Villalpando and Carlos Torres-Torres
Ultrasonic Influence on Plasmonic Effects Exhibited by Photoactive Bimetallic Au-Pt Nanoparticles Suspended in Ethanol
Reprinted from: *Materials* 2019, 12, 1791, doi:10.3390/ma12111791 91

Kenji Kinashi, Isana Ozeki, Ikumi Nakanishi, Wataru Sakai and Naoto Tsutsumi
Holographic Performance of Azo-Carbazole Dye-Doped UP Resin Films Using a Dyeing Process
Reprinted from: *Materials* 2019, 12, 945, doi:10.3390/ma12060945 107

Algirdas Lazauskas, Dalius Jucius, Valentinas Baltrušaitis, Rimantas Gudaitis, Igoris Prosyčevas, Brigita Abakevičienė, Asta Guobienė, Mindaugas Andrulevičius and Viktoras Grigaliūnas
Shape-Memory Assisted Scratch-Healing of Transparent Thiol-Ene Coatings
Reprinted from: *Materials* **2019**, *12*, 482, doi:10.3390/ma12030482 . . . . **119**

Akio Ikesue, Yan Lin Aung, Shinji Makikawa and Akira Yahagi
Total Performance of Magneto-Optical Ceramics with a Bixbyite Structure
Reprinted from: *Materials* **2019**, *12*, 421, doi:10.3390/ma12030421 . . . . **133**

Luca Oggioni, Giorgio Pariani, Frédéric Zamkotsian, Chiara Bertarelli and Andrea Bianco
Holography with Photochromic Diarylethenes
Reprinted from: *Materials* **2019**, *12*, 2810, doi:10.3390/ma12172810 . . . . **145**

About the Editor

Pierre-Alexandre Blanche currently holds the position of Research Professor at the Wyant College of Optical Sciences at the University of Arizona (Tucson, AZ, USA). Dr. Blanche has published more than 60 scientific articles in high-ranking international journals, has participated to more than 60 conferences, is the author of 12 books or book chapters, and has been granted 8 patents with 3 more pending. His fields of interest include diffraction optics, holography, 3D and see-through displays (AR/HUD), as well as non-linear and photonic materials.

Editorial

Special Issue: Photoactive Materials: Synthesis, Applications and Technology

Pierre-Alexandre Blanche

Wyant College of Optical Sciences, University of Arizona, 1630 E. University Blvd., Tucson, AZ 85721, USA; pablanche@optics.arizona.edu

Received: 22 January 2021; Accepted: 25 January 2021; Published: 27 January 2021

The science of light–matter interaction is one of the greatest accomplishment of the past 100 years. The materials that perform useful operations by collecting light or generate light from an outside stimulus are at the center of a multitude of technologies that has permeated our daily life. Everyday we rely on quantum well laser for our telecommunication, organic light emitting diode for our displays, complementary metal–oxide–semiconductor for our camera detectors, and of course a plethora of new photovoltaic cells that harvest the sunlight to satisfy our energy need.

When editing this Special Issue of *Materials* on photoactive materials, I envision collecting articles that covers a large area from photonics, to plasmonic and holography. I wanted to showcase articles that not only demonstrated new chemical synthesis, but also gave the readers a fresh view on recent applications and technologies. I was extremely pleased by how well this call for paper was received by the community, as this Special Issue presents 11 original articles and one review, all of very high quality.

In Jung et al. [1], the authors present the synthesis of hexagonal boron nitride semiconductor doped with cerium ions. Their experiment shows how the aneling of the material impacts its fluorescence emission under UV lamp. The deep blue emission intensity was dramatically enhanced when the re-heating temperature was increased. Among other uses, this material can find application as an anti-counterfeiting ink since the marking can only be identified under UV light.

The magneto-optic effect can be used for the detection of biologic phenomena such as heart beat or brain activity. Because the field generated by these events are extremely small, synthesizing materials with a large Verdet coefficient is of major importance. This importance is demonstrated by the inclusion of three articles relevant to this topic in this Special Issue [2–4].

In their manuscript, Kotov et al. [2], investigated for the first time the effects of the presence of a thin protective Bi_2O_3 layers on the magneto-optic properties of ultrathin highly bismuth-substituted dysprosium iron garnet layers. Their results showed a 2.7 times signal improvement with the protective oxide layer than without it.

Zhao and Li [3] took a different approach by doping sodium iron hexafluoride (Na_3FeF_6) particles with monodispersed terbium ions (Tb^{3+}). The synthesis was obtained by a relatively easy hydrothermal process. When measuring the magnetization according to the temperature and the external magnetic field, the authors confirmed that the Na_3FeF_6 : Tb particles were paramagnetic with a high magnetic moment.

By contrast, Ikesue et al. [4] investigated for the first time the magneto optical properties of a high-quality Bixbyite ceramics structure. They found out that the performances of these ceramics were far superior to those of commercial TGG ($Tb_3Ga_5O_{12}$) crystal, which is regarded as one of the highest class of Faraday rotator material. In particular, the Verdet constant of Tb_2O_3 (when x = 1.0) ceramic was the largest with value up to 154 rad T^{-1} m^{-1}, in the wavelength range of 633 to 1064 nm. In addition, the laser damage threshold of this ceramic was 18 J/cm^2, which is 1.8 times larger than that of TGG.

In their paper, Srinoi et al. [5], present the successful synthesis of hollow gold–silver nanoshells coated with silica shells of varying thicknesses. This was achieved by tuning the concentration of (3-aminopropyl)trimethoxysilane and sodium silicate solutions. More importantly, the authors demonstrated that none of the SiO_2 shells had detrimental effects on the localized surface plasmon resonance peak of the gold–silver nanoshells. This result is promising for the use of that type of material in plasmon-enhanced photocatalytic applications such as water-splitting reaction to directly convert sunlight into hydrogen and oxygen gases.

In a second paper regarding the optical properties of nano-particles [6], Hurtado-Aviles et al. studied the influence of an ultrasonic stimulus on the plasmonic resonance of Au–Pt nano-particles in an ethanol suspension. The presence of the ultrasonic waves in the suspension prevented any agglomerations of the nano-particles, and it was found that the light absorption associated with surface plasmon resonance was modified by the presence of the ultrasonic wave. Ultrasound interactions together to nonlinear optical phenomena in nanofluids is a promising field for applications ranging from the modulating quantum signals to use as sensors and as acousto-optic devices.

Heavy metal elements such as lead, arsenic and mercury are highly toxic for vertebrae, including human beings. Their detection in sub ppm concentration is extremely important to guarantee food and water safety. In Kurshano et al. [7], authors are considering an optical method for detecting heavy metal ions using colloidal luminescent semiconductor quantum dots. The authors combined the magnetic properties of Fe_3O_4 together with the photoluminescent properties of quantum dots of $AgInS_2/ZnS$ to detect metal ion concentration down to 0.01 pmm by measuring the quenching of the photoluminescence of their sensor.

Lembrikov, Ianetz, and Ben-Ezra are presenting a theoretical study of the nonlinear optical phenomena in a silicon waveguide with a smectic A liquid crystals (SLAC) core [8]. Authors have calculated the TM and TE modes in such a strongly anisotropic waveguide and have shown that the nonlinearity is related to the smectic layer normal displacement. They found that this effect is especially strong for the counter-propagating TM modes. By evaluating the pumping and signal TM mode slowly varying amplitudes and phases, they showed that the gain has a maximum value in the resonant case when the TM mode frequency difference is equal to the second sound frequency.

Scratch resistant coating would be of great advantage for optoelectronic components. In their study [9], Lazauskas et al. investigated the properties of transparent photopolymerizable thiol-ene coatings. These coatings exhibited high optical transparency and shape-memory that assisted scratch-healing properties. The total strain recovery ratio for the polymer were found to be up to 97% after thermal treatment. The crosslinked polymer network was also capable of initiating scratch recovery at ambient temperature.

Three articles are covering the field of holography. In Kinashi et al., the dynamic holographic recording in azo-carbazole doped polyester resin is investigated [10]. The dye-doped resin film exhibited a diffraction efficiency up to 0.23% and a response time of 5.9 s when illuminated with laser light. The dyeing process presented in the article is using aqueous solutions and offers some potential advantage for the fabrication of large-sized holographic devices as well as the manufacturing of photonic devices based on any polymer film containing organic dye.

In Blanche, Mahamat, and Buoye [11], the authors have measured some of the thermal properties of Bayfol HX200 photopolymer. This photopolymer is used as an holographic material in a variety of applications and it is important to understand the impact of temperature on the optical properties of the material. Authors found that the material as well as the hologram recorded within can sustain temperature up to 160 °C. A linear coefficient of thermal expansion (CTE) of 384×10^{-6} K^{-1} was calculated by measuring the spectral shift of a reflection hologram depending on temperature. This shows how temperature can dramatically affect the spectral response of holograms and how these measurements can be used to predict their behavior.

Finally, Oggioni et al. are presenting a review of holography using photochromic materials and, more specifically, diarylethenes dyes [12]. This re-writable class of material shows the complex modulation of their refractive index, meaning they are suitable for both amplitude and phase holograms. In addition, they and are self developing since they do not require any post processing treatment to obtain the final hologram. A combination of a kinetic model and experimental UV-vis data made possible the development of a computational tool to predict and optimize the performances of the material. The recording of both binary and grayscales holograms are presented using either a mask approach or the use of a DMD chip for direct laser writing.

I would like to thank all the authors that participated in this Special Issue, as well as the people I interacted with at MDPI for their help with the editorial process.

Funding: This research received no external funding.

Conflicts of Interest: The author declares no conflict of interest.

References

1. Jung, J.Y.; Kim, J.; Kim, Y.D.; Kim, Y.K.; Cha, H.R.; Lee, J.G.; Son, C.S.; Hwang, D. Enhanced Crystallinity and Luminescence Characteristics of Hexagonal Boron Nitride Doped with Cerium Ions According to Tempering Temperatures. *Materials* **2021**, *14*, 193. [CrossRef] [PubMed]
2. Kotov, V.; Nur-E-Alam, M.; Vasiliev, M.; Alameh, K.; Balabanov, D.; Burkov, V. Enhanced Magneto-Optic Properties in Sputtered Bi- Containing Ferrite Garnet Thin Films Fabricated Using Oxygen Plasma Treatment and Metal Oxide Protective Layers. *Materials* **2020**, *13*, 5113. [CrossRef] [PubMed]
3. Zhao, Z.; Li, X. Hydrothermal Synthesis and Optical Properties of Magneto-Optical Na3FeF6:Tb3+ Octahedral Particles. *Materials* **2020**, *13*, 320. [CrossRef] [PubMed]
4. Ikesue, A.; Aung, Y.; Makikawa, S.; Yahagi, A. Total Performance of Magneto-Optical Ceramics with a Bixbyite Structure. *Materials* **2019**, *12*, 421. [CrossRef] [PubMed]
5. Srinoi, P.; Marquez, M.; Lee, T.C.; Lee, T. Hollow Gold-Silver Nanoshells Coated with Ultrathin SiO2 Shells for Plasmon-Enhanced Photocatalytic Applications. *Materials* **2020**, *13*, 4967. [CrossRef] [PubMed]
6. Hurtado-Aviles, E.A.; Torres, J.A.; Trejo-Valdez, M.; Torres-SanMiguel, C.R.; Villalpando, I.; Torres-Torres, C. Ultrasonic Influence on Plasmonic Effects Exhibited by Photoactive Bimetallic Au-Pt Nanoparticles Suspended in Ethanol. *Materials* **2019**, *12*, 1791. [CrossRef] [PubMed]
7. Kurshanov, D.; Khavlyuk, P.; Baranov, M.; Dubavik, A.; Rybin, A.; Fedorov, A.; Baranov, A. Magneto-Fluorescent Hybrid Sensor CaCO3-Fe3O4-AgInS2/ZnS for the Detection of Heavy Metal Ions in Aqueous Media. *Materials* **2020**, *13*, 4373. [CrossRef] [PubMed]
8. Lembrikov, B.I.; Ianetz, D.; Ben-Ezra, Y. Nonlinear Optical Phenomena in a Silicon-Smectic A Liquid Crystal (SALC) Waveguide. *Materials* **2019**, *12*, 2086. [CrossRef] [PubMed]
9. Lazauskas, A.; Jucius, D.; Baltrušaitis, V.; Gudaitis, R.; Prosyčevas, I.; Abakevičienė, B.; Guobienė, A.; Andrulevičius, M.; Grigaliūnas, V. Shape-Memory Assisted Scratch-Healing of Transparent Thiol-Ene Coatings. *Materials* **2019**, *12*, 482. [CrossRef] [PubMed]
10. Kinashi, K.; Ozeki, I.; Nakanishi, I.; Sakai, W.; Tsutsumi, N. Holographic Performance of Azo-Carbazole Dye-Doped UP Resin Films Using a Dyeing Process. *Materials* **2019**, *12*, 945. [CrossRef] [PubMed]
11. Blanche, P.A.; Mahamat, A.H.; Buoye, E. Thermal Properties of Bayfol®HX200 Photopolymer. *Materials* **2020**, *13*, 5498. [CrossRef] [PubMed]
12. Oggioni, L.; Pariani, G.; Zamkotsian, F.; Bertarelli, C.; Bianco, A. Holography with Photochromic Diarylethenes. *Materials* **2019**, *12*, 2810. [CrossRef] [PubMed]

Publisher's Note: MDPI stays neutral with regard to jurisdictional claims in published maps and institutional affiliations.

© 2021 by the author. Licensee MDPI, Basel, Switzerland. This article is an open access article distributed under the terms and conditions of the Creative Commons Attribution (CC BY) license (http://creativecommons.org/licenses/by/4.0/).

Article

Enhanced Crystallinity and Luminescence Characteristics of Hexagonal Boron Nitride Doped with Cerium Ions According to Tempering Temperatures

Jae Yong Jung [1], Juna Kim [2], Yang Do Kim [2], Young-Kuk Kim [3], Hee-Ryoung Cha [3], Jung-Goo Lee [3], Chang Sik Son [1,*] and Donghyun Hwang [1,*]

1. Division of Materials Science and Engineering, Silla University, Busan 46958, Korea; eayoung21@naver.com
2. School of Materials Science and Engineering, Pusan National University, Busan 46241, Korea; kja6037@pusan.ac.kr (J.K.); yangdo@pusan.ac.kr (Y.D.K.)
3. Powder & Ceramics Division, Korea Institute of Materials Science, Changwon 51508, Korea; voice21@kims.re.kr (Y.-K.K.); h.cha@kims.re.kr (H.-R.C.); jglee36@kims.re.kr (J.-G.L)
* Correspondence: csson@silla.ac.kr (C.S.S.); dhhwang@silla.ac.kr (D.H.); Tel.: +82-10-3570-5726 (C.S.S.); +82-10-3156-4055 (D.H.)

Received: 3 December 2020; Accepted: 29 December 2020; Published: 3 January 2021

Abstract: Hexagonal boron nitride was synthesized by pyrolysis using boric acid and melamine. At this time, to impart luminescence, rare earth cerium ions were added to synthesize hexagonal boron nitride nanophosphor particles exhibiting deep blue emission. To investigate the changes in crystallinity and luminescence according to the re-heating temperature, samples which had been subjected to pyrolysis at 900 °C were subjected to re-heating from 1100 °C to 1400 °C. Crystallinity and luminescence were enhanced according to changes in the reheating temperature. The synthesized cerium ion-doped hexagonal boron nitride nanoparticle phosphor was applied to the anti-counterfeiting field to prepare an ink that can only be identified under UV light.

Keywords: hexagonal boron nitride; photoluminescence; cerium; anti-counterfeiting; crystals

1. Introduction

Hexagonal boron nitride (h-BN) is an important material with excellent properties, such as a wide band gap (4.4~6.0 eV), high electrical insulation, low dielectric constant, high temperature stability and large-scale oxidation cross-sectional area, and thermal neutrons [1–3]. All these features have made h-BN a promising material in the aviation industry. Moreover, it has many uses in microelectronic mechanical systems (MEMs), biomedicine, fireproofing, laser devices, solid-state neutron detectors, lubricants and electrical insulators [4–6]. Among the features of h-BN that make it suitable for such varied uses, the wide band gap allows it to be used in applications that require a material with unique luminescence characteristics.

Since the first observation of intense far-ultraviolet (UV) excitation emission, the unique optical and fluorescent properties of h-BN have attracted special attention in the past ten years, making this material a candidate for new light-emitting devices, and it is expected to be used as a future material for photovoltaic applications [7–9]. In addition, the band gap energy of h-BN can be adjusted by designing a super lattice, doping, recombination of organic functional groups and surface functionalization. Doping with rare earth (RE) ions is the most widely known method of tuning the band gap energy of h-BN. Continued interest in rare earth-doped nitride-based materials appears to be increasing, especially for h-BN, partly because of the discovery that the thermal quenching of luminescence decreases as the bandgap of the host materials increases [10,11]. In particular, metal nitrides or oxynitrides containing activator ions are often used as color-converting phosphors for white light-emitting diodes (wLEDs), because they have excellent visible light emission and less thermal quenching. However, most nitride-based host materials used in phosphors are ternary or quaternary nitride semiconductors. Binary nitride semiconductors, such as aluminum nitride (AlN), gallium nitride (GaN) and BN are rarely used as host materials for luminescent activators [12–14]. Steckl et al. proposed a diode that emits green light in the visible region by doping erbium ions using GaN as a host material [15]. In a study by Jadwisienczak et al., AlN thin film doped with terbium ions by sputtering was grown on a silicon wafer and a light-emitting device in the visible region was produced by a re-heating process [16]. Kim et al. investigated the changes in the luminescence of deep blue with various doping concentrations of cerium ions using h-BN as a host [17]. Mauro et al. described the optical properties of dilute semiconductor materials by presenting a set of highly efficient analytical equations that focus on the evolution of the peak luminescence gain with temperature and the relationship to sample quality [18].

In this study, the crystallinity and luminescence properties of cerium ion-doped hexagonal boron nitride (h-BN) according to the change of reheating temperature were investigated by X-ray diffraction, transmission electron microscopy, Raman spectroscopy, photoluminescence and photoluminescence excitation analysis.

2. Materials and Methods

2.1. Synthesis of Hexagonal Boron Nitride Nanophosphor Doped with Ce^{3+}

The experimental procedure is schematically shown in Figure 1. Hexagonal boron nitride nanophosphors were synthesized from boric acid (H_3BO_3, Sigma-Aldrich, St. Louis, MO, USA, ≥98.5%) and melamine ($C_3H_6N_6$, Sigma-Aldrich, 99%). First, 7 mmol of H_3BO_3 and 1 mmol of $C_3H_6N_6$ were dissolved in 150 mL deionized (D.I) water. Cerium 0.05 mmol was also incorporated by dissolution of cerium nitrate ($Ce(NO_3)_3 \cdot 6H_2O$, Sigma-Aldrich, St. Louis, MO, USA, 99.999%). These materials were completely dissolved in D.I water, evaporated at 120 °C while stirring at 500 rpm to obtain the precursor, and then dried at 80 °C for 24 h. The dried precursor underwent pyrolysis in an alumina tubular furnace (AJEON FURNACE, Namyangju, Korea) at 900 °C under nitrogen atmosphere. After heat treatment, the sample was rinsed with D.I water to remove any remaining unreacted material. When the powder was collected by agglomeration in the direction of gravity using a centrifuge, the solution was discarded, and the recovered powder specimen was dried at 80 °C for 12 h. The samples of h-BN synthesized at 900 °C were subjected to re-heating at 1100, 1200, 1300 and 1400 °C for 2 h under nitrogen atmosphere to investigate their crystallinity and luminescence in relation to the tempering temperature.

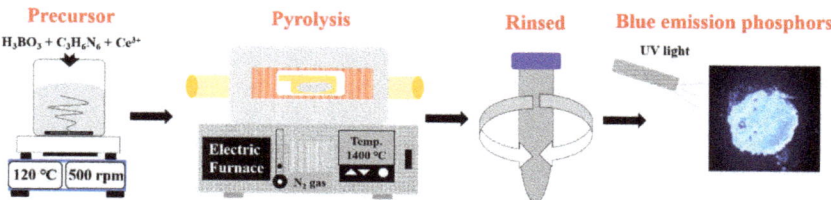

Figure 1. Schematic of preparation of the Ce^{3+}-doped h-BN nanophosphors.

2.2. Characterization of h-BN Nanophosphors

X-ray powder diffraction (XRD, Rigaku, Ultima IV, Tokyo, Japan) analysis was performed to investigate the crystallinity of the samples in relation to various temperatures with Cu-Kα radiation (0.15406 nm) generated at 40 kV and 20 mA. The measurement range was performed from 20 to 80 °C and the step scan was performed for 4 s at 0.02 degree intervals. Raman spectra were recorded using a dispersive Raman spectrometer (LabRam-HR 800, Horiba Jobin-Yvon, Longjumeau, France) equipped with a microscope and a 522-nm laser as the excitation source. In addition, the specimen was made into a 1 mm × 1 mm cylindrical pellet and measured in the range of 700 to 2300 after focusing at 1800 gv magnification and focus using the E2g mode of the optical lens. The photoluminescence (PL) of the samples was measured and photoluminescence excitation spectroscopy (PLE) was conducted with a fluorescence spectrometer (FP-6500, JASCO, Tokyo, Japan) equipped with a xenon (Xe) flash lamp. At this time, the supplied energy of the spectral lamp was measured with 200 photomultiplier tube modules (PMT). The PL measurement range was from 250 to 800 nm, and the PLE was from 200 to 550 nm. The morphology and crystallinity of the nanophosphors were observed by transmission electron microscopy (TEM, JEM 2100F, JEOL, Tokyo, Japan). The element distribution was analyzed by energy dispersive X-ray spectroscopy (EDX, X-Max 150, Oxford Instruments, Abingdon, UK) to confirm the presence of rare earths in the samples. The resolution was about 129 eV and the analysis time was X-ray exposure for 2 min for each specimen. At this time, the components were compared with B, N, Ce and C for the component detection analysis of the synthesized parent and the doped material. Spectrometry (V-570, JASCO, Tokyo, Japan) was conducted to investigate the transmittance of the bare glass and thin film-coated glass substrate. The analysis range was from 250 to 800 nm, and analysis was performed using an integrating sphere.

2.3. Applied Anti-Counterfeiting and Fingerprinting

Anti-counterfeiting inks prepared using Ce^{3+}-doped h-BN nanophosphors 1 wt% were dispersed in an aqueous solution containing 10 wt% of polyvinylpyrrolidone (PVP, M.W. = 14,000). The solution was spin coated on glass at 2000 rpm and banknotes were painted with a brush and then dried at 80 °C for an hour. The thin film coated on glass was photographed under UV light. Fingerprints were marked on glass substrates. Then Ce^{3+}-doped h-BN nanophosphor powders were applied to the glass substrate surface and latent fingerprints on the surfaces were carefully wiped off. The latent fingerprints coated with the nanophosphors were developed using a UV lamp, and the appearance of the fingerprints was confirmed by photography.

3. Results and Discussion

3.1. Crystallinity and Morphology of Ce^{3+}-Doped h-BN Nanophosphors

Ce^{3+}-doped h-BN nanophosphors were synthesized by pyrolysis of a precursor in the form of a chemical compound prepared from boric acid, melamine and cerium nitrate at 900 °C under a nitrogen atmosphere. The samples were re-heated at 1100, 1200, 1300 and 1400 °C and their crystallinity was investigated by analysis of their XRD patterns as shown in Figure 2a. In all of the synthesized samples, the peak of the (002) phase matched the Joint Committee on Powder Diffraction Standards (JCPDS 34-0421) and the main diffraction peak of hexagonal boron nitride crystalline was observed in the XRD patterns. In the case of the sample synthesized at 900 °C, the shape of the (002) phase, which was the main diffraction peak, was slightly broad and showed a relatively weak signal. However, as the re-heating temperature increased, the XRD signal of the main diffraction peak (002) phase changed strongly and clearly. These changes were not the perfect form of h-BN for a sample synthesized at a relatively low temperature. Rather, the formed phase was turbo-stratic boron nitride (t-BN), which has the same crystal structure as h-BN but has hexagonal layers stacked and randomly rotated along the c-axis identified [19]. As the re-heating temperature was increased, the profiles of the diffraction peaks became clearer and their full width at half maximum (FWHM) narrowed (Figure 2b, black symbol); i.e., the crystallinity of t-BN formed by the pyrolysis of the sample synthesized at 900 °C improved as the re-heating temperature increased. The interplanar spacing of the (002) peak in the diffraction patterns of the Ce^{3+}-doped h-BN nanophosphors gradually increased as the re-heating temperature increased.

The FWHM of the samples decreased (Figure 2b, black symbol) and the interplanar spacing of the (002) phase increased (Figure 2b, red symbol).

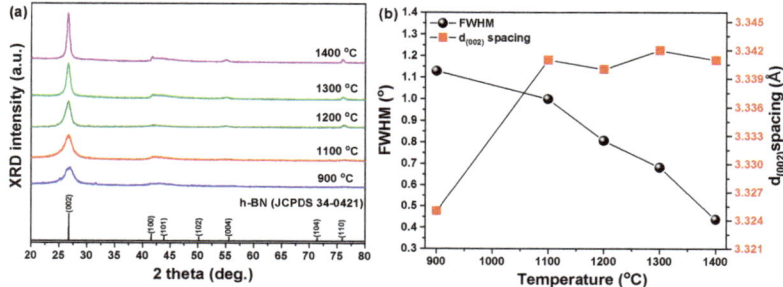

Figure 2. (a) X-ray powder diffraction (XRD) patterns, (b) full width at half maximum (FWHM) and interplanar spacing of (002) phase according to 900, 1100, 1200, 1300 and 1400 °C temperatures.

Increased interplanar spacing and decreased FWHM of (002) peaks are commonly found during crystallization of h-BN [20]. We can see that the crystallinity of h-BN is improved with increased re-heating temperature. For these assumptions to be valid, Raman analysis expressed as the lattice frequency of an intrinsic constant was performed, and the results are shown in Figure 3. The Ce^{3+}-doped h-BN nanophosphors' position converts the E_{2g} mode near 1363 cm^{-1} of Raman spectra to low frequencies. Because the frequency of the Raman spectrum mode is inversely proportional to the square root of the constituent atomic mass [21], the transition to the lower frequency of the Raman spectrum means that heavy Ce atoms are incorporated into the BN lattice of light elements.

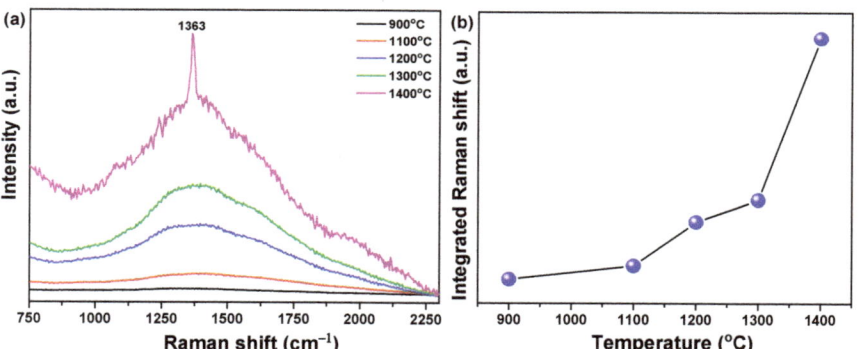

Figure 3. (a) Raman shift, (b) change of integrated and FWHM (inset) of Raman shift according to 900, 1100, 1200, 1300 and 1400 °C temperatures.

In addition, according to the re-heating temperature, the increased integrated area and decreased FWHM of the Raman shift indicate enhanced crystallinity, which is consistent with the XRD data. In the case of the specimen pyrolyzed at 900 °C and reheated at 1100 °C, Raman signals were hardly observed. However, when the reheating temperature was increased to 1200 °C or higher, the broad Raman signal strength became stronger and a distinct and strong Raman signal was observed at 1400 °C. This tendency is thought to be a result of the increase in crystallinity with increased reheating temperature.

Transmission electron microscopy was performed to observe the shape and morphology of the synthesized Ce^{3+}-doped h-BN nanophosphors as shown in Figure 4. Figure 4a shows a sample pyrolyzed at 900 °C and the shape of the nanoparticles is unclear. When the lattice constant (0.332 nm) was profiled with high-resolution, it was found to be close to the XRD result. In addition, carbon was observed (Figure 4c) throughout the sample; this was unreacted and unburned material residue due to the relatively low heat-treatment temperature. However, the sample synthesized at 900 °C and re-heated at 1400 °C showed enhanced crystallinity according to the XRD results and showed a distinct elliptical plate shape of about 20-nm in the TEM image. The lattice constant $d_{(002)}$ observed with high-resolution also decreased (0.334 nm) from the XRD data, and components of B, N and Ce were detected by EDX analysis (Figure 4d), confirming that Ce ions were doped in the h-BN lattice. The XRD and TEM results showed that, as the re-heating temperature increased, the enhanced crystallinity and the size of the particles increased [22]. It is thought to be an important process for synthesizing h-BN.

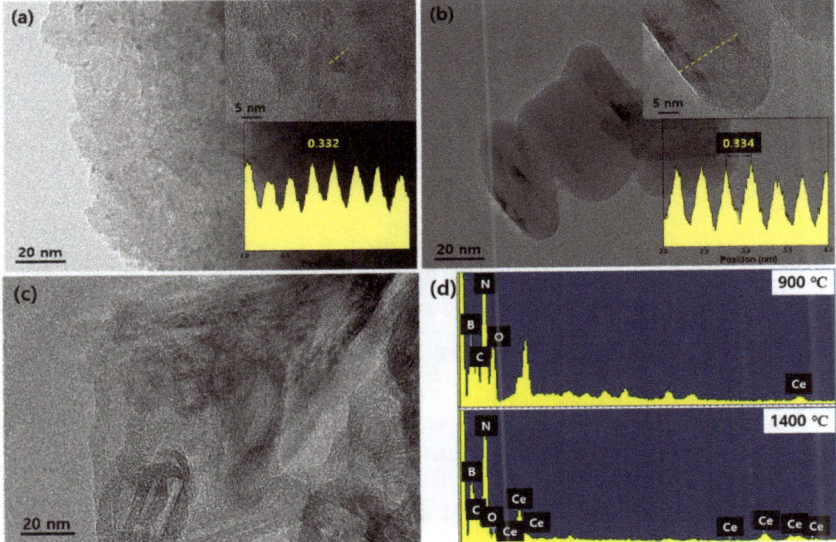

Figure 4. (**a**) Transmission electron microscopy (TEM) image of pyrolysis at 900 °C of Ce^{3+}-dope h-BN nanophosphors (inset up; high-resolution, inset down; line profile), (**b**) TEM image of re-heating at 1400 °C of Ce^{3+}-dope h-BN nanophosphors (inset up; high-resolution, inset down; line profile), (**c**) TEM image of observed remain carbon in pyrolysis at 900 °C and (**d**) X-ray spectroscopy (EDX) data of pyrolysis at different tempering temperatures (element identification: B, C, O, N and Ce).

3.2. Luminescence of Ce^{3+}-Doped h-BN Nanophosphors

Figure 5 shows the changes in luminescence according to the re-heating temperature of the Ce^{3+}-doped h-BN synthesized at 900 °C. When the excitation wavelength was controlled at 304 nm, the photoluminescence (PL) wavelength showed deep-blue emission at 396 nm. The PL intensity was enhanced remarkably as the re-heating temperature increased as shown in Figure 5b. The increase in the PL intensity of the sample with increased re-heating temperature may be attributed to the substitutional incorporation of luminescent Ce^{3+} ions into the enhanced crystallinity lattice, as revealed in the previous section. The photoluminescence excitation (PLE) spectra of Ce^{3+}-doped h-BN nanophosphors irradiated for emission at 396 nm showed an asymmetric curve centered at 304 nm. This is actually a mirror image of the PL spectra. This has been reported for organic dyes in the emission spectra [23]. Mirror symmetry is usually found for defect centers with weak Jahn-Teller interactions [24], which can be attributed to the substitution of Ce^{3+} ions for B atoms in the hexagonal boron nitride framework, as shown in Raman spectroscopy. The broad PL spectrum of cerium-doped nitride materials is usually attributed to the 4f-5d excitation of Ce^{3+} ions [25].

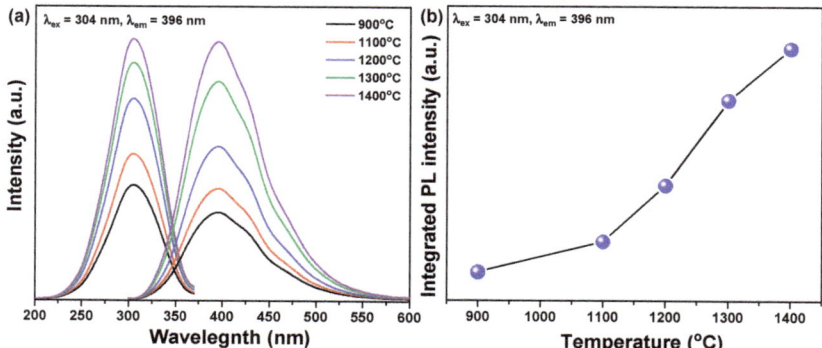

Figure 5. (a) Photoluminescence excitation (PLE) and photoluminescence (PL) spectra and (b) integrated PL intensity of Ce^{3+}-doped h-BN nanophosphors in relation to re-heating temperatures of 900, 1100, 1200, 1300 and 1400 °C.

Because 5d electrons contribute to chemical bonds, the PL intensity and emission band are highly dependent on the chemical environment surrounding the Ce^{3+} ions [26]. As a result, the PL spectrum was highly dependent on the crystallinity of the host material. Interestingly, further increase in the re-heating temperature enhanced the crystallinity of Ce^{3+}-doped h-BN nanophosphors.

3.3. Anti-Counterfeiting Application of Ce^{3+}-Doped h-BN Nanophosphors

The large intrinsic bandgap (≥5 eV) of BN and the plate-like particle shape allow rare earth-doped boron nitride to be deposited as transparent phosphors on a flat surface. Figure 6a shows a Ce^{3+}-doped h-BN nanophosphors thin film deposited on a glass substrate.

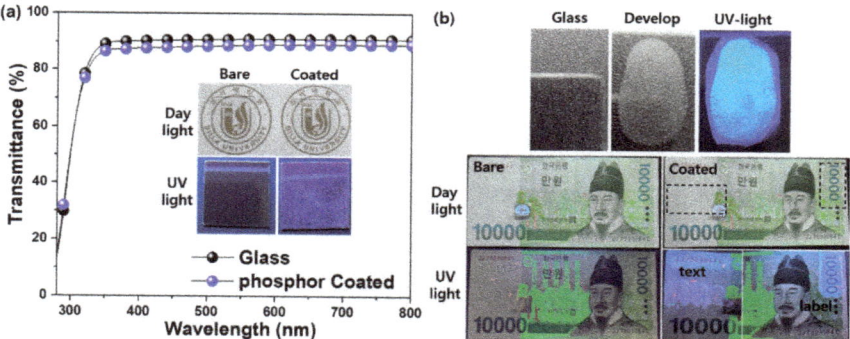

Figure 6. (a) Transmittance of Ce^{3+}-doped h-BN nanophosphors coated on glass and (b) images of Ce^{3+}-doped h-BN nanophosphors fingerprinting development on glass substrate (up) and painted on Korean bank notes (down); used samples were re-heated at 1400 °C.

The transmittance of the glass substrate decreased slightly after coating with the phosphor thin film with a colloidal solution of Ce^{3+}-doped h-BN nanophosphors. In daylight, it was difficult to distinguish between the bare glass and the phosphor-coated glass with the naked eye due to the transparency of the phosphor thin film. However, when irradiated with UV light, strong blue light emission can be seen from the phosphor film. The images of fingerprinting development are shown in the upper part of Figure 6b. The bare, donor and powdered images used Ce^{3+}-doped h-BN nanophosphors re-heated at 1400 °C. The emission image was taken under UV light. The fingerprint was obtained from a person whose fingerprint has a whorl loop. The bare image was blurred in daylight. The powdered images produced by using prepared nanophosphors were eye-catching, because these particles were attached to the moisture component of the fingerprint. Under UV light, the blue emission confirmed that the contrast and resolution of the fingerprint had been improved. A Ce^{3+}-doped h-BN nanophosphor colloidal solution was painted on a Korean bank note surface as shown in the lower images in Figure 6b. In daylight, it is difficult to distinguish between bare bank notes and those coated with the phosphor thin film. However, the phosphor-coated bank notes showed an intense blue emission text 'Silla' and label under UV light. Due to its light transmission and visible light emission under UV irradiation, this is hidden under normal conditions and can be recognized by the naked eye under UV radiation. Labels and text composed of h-BN based nanophosphors can be effectively hidden in daylight but can be easily identified under UV light, which is essential for a variety of anti-counterfeiting applications.

4. Conclusions

Hexagonal boron nitride doped with cerium ions was successfully synthesized by pyrolysis and re-heating of the precursor prepared from a chemical mixture of boric acid, melamine and cerium nitrate. Deep blue emission was detected at 396 nm under excitation with 304 nm from Ce^{3+}-doped h-BN nanophosphors. The deep blue emission intensity was notably enhanced by increasing re-heating temperature. Increased re-heating temperature resulted in improved crystallinity and luminescence. It also affected particle growth. The deep blue emission is attributed to the transition of Ce^{3+} ions from the 5d level crystal field component to the 4f ground state. In the case of the specimen synthesized at 900 °C, the results of XRD patterns and Raman spectrum inferior crystallinity due to relatively low temperature were shown. However, as the reheating treatment temperature increased, the crystallinity improved, which is thought to be the effect of recrystallization caused by heat energy supplied from the outside. In addition, in the TEM image, particles with clear crystals were observed at 1400 °C, where the reheating treatment temperature was the highest and, as a result of component analysis, B, N and Ce were identified. This improvement in crystallinity has resulted in an increase in luminescence properties. The solution-based coating of Ce^{3+}-doped h-BN nanophosphors was applied to various substrates; the films showed excellent transparency and luminescence with a strong visible blue color. The hiding and easy identification of nanophosphors under normal conditions demonstrates the feasibility of using rare earth-doped hexagonal boron nitride in anti-counterfeiting inks in a variety of applications.

Author Contributions: Conceptualization, J.Y.J., D.H. and Y.-K.K.; methodology, J.Y.J., Y.-K.K., C.S.S., Y.D.K. and J.-G.L.; formal analysis, J.Y.J., H.-R.C., J.K. and D.H.; investigation, J.Y.J., J.-G.L., H.-R.C. and J.K.; writing—original draft preparation, J.Y.J.; writing—review and editing, J.Y.J., Y.D.K., D.H. and C.S.S.; visualization, J.Y.J., and Y.D.K.; supervision, D.H.; project administration, C.S.S. All authors have read and agreed to the published version of the manuscript.

Funding: This research received no external funding.

Institutional Review Board Statement: Not applicable.

Informed Consent Statement: Informed consent was obtained from all subjects involved in the study.

Data Availability Statement: The data presented in this study are available in the database of the authors at the Faculty of Materials Science and Engineering.

Acknowledgments: This research was supported by Basic Science Research Program through the National Research Foundation of Korea (NRF) funded by the Ministry of Education (No. NRF-2019R1A6A3A01095400); and by the National Research Foundation of Korea (NRF) grant funded by the Korea government (MSIT) (No. NRF-2018R1A5A1025594); and by the Korea Institute of Energy Technology Evaluation and Planning (KETEP) and the Ministry of Trade, Industry & Energy (MOTIE) of the Republic of Korea (No. 20193010014850).

Conflicts of Interest: The authors declare no conflict of interest.

References

1. Jiang, H.X.; Lin, J.Y. Review-Hexagonal Boron Nitride Epilayers: Growth, Optical Properties and Device Applications. *ECS J. Solid State Sci. Technol.* **2016**, *6*, 3012–3021. [CrossRef]
2. Favennec, P.N.; L'Haridon, H.; Salvi, M.; Moutonnet, D.; Le Guillou, Y. Luminescence of erbium implanted in various semiconductors: IV, III-V and II-VI materials. *Electron. Lett.* **1989**, *25*, 718–719. [CrossRef]
3. Dean, C.R.; Young, A.F.; Meric, I.; Lee, C.; Wang, L.; Sorgenfrei, S.; Watanabe, K.; Taniguchi, T.; Kim, P.; Shepard, K.L.; et al. Boron nitride substrates for high-quality graphene electronics. *Nat. Nanotechnol.* **2010**, *5*, 722–726. [CrossRef] [PubMed]
4. Huang, B.; Cao, X.K.; Jiang, H.X.; Lin, J.Y.; Wei, S. Origin of the significantly enhanced optical transitions in layered boron nitride. *Phys. Rev. B Condens. Matter Mater. Phys.* **2012**, *86*, 155202. [CrossRef]
5. Song, L.; Ci, L.; Lu, H.; Sorokin, P.B.; Jin, C.; Ni, J.; Kvashnin, A.G.; Kvashnin, D.G.; Lou, J.; Yakobson, B.I.; et al. Large Scale Growth and Characterization of Atomic Hexagonal Boron Nitride Layers. *Nano. Lett.* **2010**, *10*, 3209–3215. [CrossRef]
6. Sugino, T.; Tanioka, K.; Kawasaki, S.; Shirafuji, J. Characterization and Field Emission of Sulfur-Doped Boron Nitride Synthesized by Plasma-Assisted Chemical Vapor Deposition. *Jpn. J. Appl. Phys.* **1997**, *36*, 463–466. [CrossRef]
7. Kubota, Y.; Watanabe, K.; Tsuda, O.; Taniguchi, T. Deep Ultraviolet Light-Emitting Hexagonal Boron Nitride Synthesized at Atmospheric Pressure. *Science (Am. Assoc. Adv. Sci.)* **2007**, *317*, 932–934. [CrossRef]
8. Oder, T.N.; Kim, K.H.; Lin, J.Y.; Jiang, H.X. III-nitride blue and ultraviolet photonic crystal light emitting diodes. *Appl. Phys. Lett.* **2004**, *84*, 466–468. [CrossRef]
9. Silly, M.G.; Jaffrennou, P.; Barjon, J.; Lauret, J.-S.; Ducastelle, F.; Loiseau, A.; Obraztsova, E.; Attal-Tretout, B.; Rosencher, E. Luminescence properties of hexagonal boron nitride: Cathodoluminescence and photoluminescence spectroscopy measurements. *Phys. Rev. B Condens. Matter Mater. Phys.* **2007**, *75*, 085205. [CrossRef]
10. Majety, S.; Cao, X.K.; Li, J.; Dahal, R.; Lin, J.Y.; Jiang, H.X. Band-edge transitions in hexagonal boron nitride epilayers. *Appl. Phys. Lett.* **2012**, *101*, 51110. [CrossRef]
11. Watanabe, K.; Taniguchi, T.; Kanda, H. Direct-bandgap properties and evidence for ultraviolet lasing of hexagonal boron nitride single crystal. *Nat. Mater.* **2004**, *3*, 404–409. [CrossRef] [PubMed]
12. Museur, L.; Kanaev, A. Near band-gap photoluminescence properties of hexagonal boron nitride. *J. Appl. Phys.* **2008**, *103*, 103520. [CrossRef]
13. Ahmad, P.; Khandaker, M.U.; Amin, Y.M.; Muhammad, N.; Khan, G.; Khan, A.S.; Numan, A.; Rehman, M.A.; Ahmed, S.M.; Khan, A. Synthesis of hexagonal boron nitride fibers within two hour annealing at 500 °C and two hour growth duration at 1000 °C. *Ceram. Int.* **2016**, *42*, 14661–14666. [CrossRef]
14. Wu, J.; Yi, L.; Zhang, L. Tuning the electronic structure, bandgap energy and photoluminescence properties of hexagonal boron nitride nanosheets via a controllable Ce^{3+} ions doping. *RSC Adv.* **2013**, *3*, 7408. [CrossRef]
15. Steckl, A.J.; Garter, M.; Birkhahn, R.; Scofield, J. Green electroluminescence from Er-doped GaN Schottky barrier diodes. *Appl. Phys. Lett.* **1998**, *73*, 2450–2452. [CrossRef]
16. Jadwisienczak, W.M.; Lozykowski, H.J.; Perjeru, F.; Chen, H.; Kordesch, M.; Brown, I.G. Luminescence of Tb ions implanted into amorphous AlN thin films grown by sputtering. *Appl. Phys. Lett.* **2000**, *76*, 3376–3378. [CrossRef]
17. Jung, J.; Baek, Y.; Lee, J.; Kim, Y.; Cho, S.; Kim, Y. The structure and luminescence of boron nitride doped with Ce ions. *Appl. Phys. A* **2018**, *124*, 1–6. [CrossRef]
18. Pereira, M.F. Analytical Expressions for Numerical Characterization of Semiconductors per Comparison with Luminescence. *Materials* **2018**, *11*, 2. [CrossRef]
19. Alkoy, S.; Toy, C.; Gönül, T.; Tekin, A. Crystallization behavior and characterization of turbostratic boron nitride. *J. Eur. Ceram. Soc.* **1997**, *17*, 1415–1422. [CrossRef]
20. Sarkar, S.; Gan, Z.; An, L.; Zhai, L. Structural Evolution of Polymer-Derived Amorphous SiBCN Ceramics at High Temperature. *J. Phys. Chem. C* **2011**, *115*, 24993–25000. [CrossRef]
21. Gorbachev, R.V.; Riaz, I.; Nair, R.R.; Jalil, R.; Britnell, L.; Belle, B.D.; Hill, E.W.; Novoselov, K.S.; Watanabe, K.; Taniguchi, T.; et al. Hunting for Monolayer Boron Nitride: Optical and Raman Signatures. *Small (Weinh. Bergstr. Ger.)* **2011**, *7*, 465–468. [CrossRef] [PubMed]

22. Wu, J.; Han, W.; Walukiewicz, W.; Ager, J.W.; Shan, W.; Haller, E.E.; Zettl, A. Raman Spectroscopy and Time-Resolved Photoluminescence of BN and $B_xC_yN_z$ Nanotubes. *Nano Lett.* **2004**, *4*, 647–650. [CrossRef]
23. Li, J.; Yuan, C.; Elias, C.; Wang, J.; Zhang, X.; Ye, G.; Huang, C.; Kuball, M.; Eda, G.; Redwing, J.M.; et al. Hexagonal Boron Nitride Single Crystal Growth from Solution with a Temperature Gradient. *Chem. Mater.* **2020**, *32*, 5066–5072. [CrossRef]
24. Wu, Y.; Chen, Y.; Wang, D.; Lee, C.; Sun, C.; Chen, T. α-(Y,Gd)FS:Ce^{3+}: A novel red-emitting fluorosulfide phosphor for solid-state lighting. *J. Mater. Chem.* **2011**, *21*, 15163. [CrossRef]
25. Chowdhury, C.; Jahiruddin, S.; Datta, A. Psuedo Jahn-Teller Distortion in Two-Dimensional Phosphorus: Origin of Black and Blue Phases of Phosphorene and Band Gap Modulation by Molecular Charge Transfer. *J. Phys. Chem. Lett.* **2016**, *7*, 1288–1297. [CrossRef]
26. Qin, X.; Liu, X.; Huang, W.; Bettinelli, M.; Liu, X. Lanthanide-Activated Phosphors Based on 4f-5d Optical Transitions: Theoretical and Experimental Aspects. *Chem. Rev.* **2017**, *5*, 4488–4527. [CrossRef]

Publisher's Note: MDPI stays neutral with regard to jurisdictional claims in published maps and institutional affiliations.

© 2021 by the authors. Licensee MDPI, Basel, Switzerland. This article is an open access article distributed under the terms and conditions of the Creative Commons Attribution (CC BY) license (http://creativecommons.org/licenses/by/4.0/).

Article

Thermal Properties of Bayfol® HX200 Photopolymer

Pierre-Alexandre Blanche [1,*], Adoum H. Mahamat [2] and Emmanuel Buoye [2]

1. College of Optical Sciences, University of Arizona, 1630 E. University Blvd, Tucson, AZ 85721, USA
2. Naval Air Systems Command/NAWCAD, Patuxent River, MD 20670, USA; adoum.mahamat@navy.mil (A.H.M.); emmanuel.buoye1@navy.mil (E.B.)
* Correspondence: pablanche@optics.arizona.edu

Received: 22 October 2020; Accepted: 26 November 2020; Published: 2 December 2020

Abstract: Bayfol® HX200 photopolymer is a holographic recording material used in a variety of applications such as a holographic combiner for a heads-up display and augmented reality, dispersive grating for spectrometers, and notch filters for Raman spectroscopy. For these systems, the thermal properties of the holographic material are extremely important to consider since temperature can affect the diffraction efficiency of the hologram as well as its spectral bandwidth and diffraction angle. These thermal variations are a consequence of the distance and geometry change of the diffraction Bragg planes recorded inside the material. Because temperatures can vary by a large margin in industrial applications (e.g., automotive industry standards require withstanding temperature up to 125 °C), it is also essential to know at which temperature the material starts to be affected by permanent damage if the temperature is raised too high. Using thermogravimetric analysis, as well as spectral measurement on samples with and without hologram, we measured that the Bayfol® HX200 material does not suffer from any permanent thermal degradation below 160 °C. From that point, a further increase in temperature induces a decrease in transmission throughout the entire visible region of the spectrum, leading to a reduced transmission for an original 82% down to 27% (including Fresnel reflection). We measured the refractive index change over the temperature range from 24 °C to 100 °C. Linear interpolation give a slope $4.5 \times 10^{-4}\ K^{-1}$ for unexposed film, with the extrapolated refractive index at 0 °C equal to $n_0 = 1.51$. This refractive index change decreases to $3 \times 10^{-4}\ K^{-1}$ when the material is fully cured with UV light, with a 0 °C refractive index equal to $n_0 = 1.495$. Spectral properties of a reflection hologram recorded at 532 nm was measured from 23 °C to 171 °C. A consistent 10 nm spectral shift increase was observed for the diffraction peak wavelength when the temperature reaches 171 °C. From these spectral measurements, we calculated a coefficient of thermal expansion (CTE) of $384 \times 10^{-6}\ K^{-1}$ by using the coupled wave theory in order to determine the increase of the Bragg plane spacing with temperature.

Keywords: photopolymer; temperature; hologram; CTE; thermal degradation; refractive index

1. Introduction

Over the past few decades, most imaging and nonimaging systems have been designed and built using conventional bulky glass- and metal-based optical elements. Those optical elements have been proved to perform well under many conditions, but they are heavy, expensive, and require long lead time to manufacture. In recent years, optical designers started to shift their focus on the design of thin, lightweight, and easy-to-manufacture optical elements using the method of holographic recording [1–3]. Typically, holographic optical elements can be recorded on several different photosensitive materials such as dichromated gelatin, silver halide, photoresist, and photopolymer [4].

As the demand for small, lightweight, and compact optical systems has grown since augmented and virtual reality technologies have entered the optics industry, interest in holographic optical elements has also rapidly increased. Holographic optical elements are cheap, easy to manufacture, and sensitive to wavelength and incidence angle; they also provide high diffraction efficiency. These optical elements diffract light through the refractive index modulation obtained through their recording and development processes [5,6].

For many applications, the thermal response of the holographic material is particularly important. The linear coefficient of thermal expansion (CTE) can have an effect on both the diffraction spectrum and diffraction angle of the hologram, changing the color and modifying the direction of the diffracted beam [7].

In the case of a holographic optical element such as a lens, the temperature can change the focal length [8]. In the case of dispersion gratings or holographic notch filters, there can be a shift in the wavelength distribution that affects their use in spectrometry or in a laser cavity [9–11]. For a grating coupler such as those used in holographic combiners for augmented reality and for a heads-up display, this could modify both the color and the field of view of the system [12–14].

It is also important to consider the thermal stability of the holographic material for integration into industrial processes and applications [15]. The holographic material and the diffractive structure contained within should be able to withstand the high temperatures encountered during thermoplastic molding and extrusion, multilayers lamination (such as in windshield and security windows), and their use in extreme environments such as defined in automotive, military, and aerospace specifications (up to +125 °C) [16,17].

Volume holograms recorded in acrylamide-based photopolymers were investigated for their operational range and reversibility over temperature range of 15–50 °C, and relative humidity of 10–80% [18–20]. This material experiences a red wavelength shift in its diffraction response with increased temperature and humidity. Liu et al. [21] investigated the spectral properties of the DCG-based holograms under different temperatures and humidity conditions. They found that the peak diffracted wavelength decreased with temperature and thermal processing time. Lin et al. [22] studied the temperature effect in PQ:PMMA photopolymer and reported enhancement of the diffraction efficiency with temperature postprocessing. SU-8, a commercial photoresin used for photolithography and surface relief holographic gratings, has been shown to shrink, soften, and even collapse at temperatures just above 100 °C. [23]. However, to our knowledge, no one has studied the thermal response of the Bayfol® HX200 photopolymer material.

Bayfol® HX200 photopolymer is a holographic recording material distributed by Covestro that has found uses in many applications such as the heads-up display, augmented reality glasses, solar concentrator, and disperser for a spectrometer [14,24,25]. Bayfol® HX200 is an acrylate-based 2-chemistry photopolymer material; its composition and chemistry are described in detail in [26]. The chemical and physical properties of this material, such as the photo-initiation process, spectral photosensitivity, refractive index modulation, and bleaching, have been characterized [26–28]. The Bayfol® HX200 film has an average refractive index of $n_o = 1.49$ and a thickness of $d = 16$ µm, and it can provide a maximum refractive index modulation of $\Delta n = 0.03$, per the manufacturer's specifications [29].

Unfortunately, little is known about its thermal response. The only available information, to our knowledge, is the fact that the Bayfol® HX200 is able to withstand the injection molding procedure with a mold temperature of 70 °C [30].

In this paper, we present a study of several thermal characteristics of the Bayfol® HX200 that are critical for its use as an holographic material in industrial processes. We measured the thermal degradation using both thermogravimetry and the transmission spectrum. Both techniques are complementary, giving the maximum temperature under which the material should be kept to ensure its proper optical operation. We quantified the variation of the refractive index as a function of temperature between 24 °C and 100 °C, for unexposed as well as for exposed materials. Knowing the refractive index allows determination of the precise optical path of the light rays inside the sample and permits the accurate optical design of systems that include this material. This is particularly important for holographic waveguides that rely on total internal reflection, which is determined by the refractive index. Finally, we are also reporting the measurement of the CTE of the material calculated from the wavelength drift experienced by a reflection hologram with temperature. The CTE and the refractive index are absolutely necessary parameters to calculate the thermal behavior of an optical system that use the Bayfol® HX200 material. Together, these two parameters determine the change in spectral and angular dispersion of holographic optical elements according to temperature, and they were not known before this study.

2. Materials and Methods

2.1. Thermal Degradation

To measure the thermal stability of the Bayfol® HX200 material (manufactured by Covestro AG, Leverkusen, Germany), we used both thermogravimetric analysis and spectroscopy.

The thermogravimetric analysis was performed on a TA Instruments (New Castle, DE, USA) TGA 550 machine at a rate of 10 °C/min. The photopolymer sample was separated from the backing film and both compounds were measured separately.

To make sure the material does not suffer any optical damage at temperatures below the weight loss recorded by the thermogravimetric experiment, we recorded the transmission spectrum depending on temperature. For this experiment, the sample was prepared as follows: a 4 × 4 cm² piece of Bayfol® HX200 material was laminated on a float glass and exposed to sunlight for 5 min until fully bleached. The material was then encapsulated with another float glass using UV curing optical glue NOA61, which can withstand a temperature of 260 °C for three hours according to the manufacturer.

The sample was mounted parallel in front of a first surface aluminum mirror and placed into a temperature-regulated oven. The front side of the oven had a transparent window so that the sample could be illuminated. We used an Oceanoptics (now Ocean Insight, Orlando, FL, USA) USB-4000 fiber fed UV-VIS spectrometer to measure the transmission spectrum of the sample in a double-pass experiment where the back mirror reflected the light back at the spetrometer input fiber. The light source was a halogen lamp that was projected on a diffuser, passing through an aperture, and collimated. The amplitude of the spectrum was calibrated using a dark measurement where the light source was turned off, and a 100% measurement where the sample was removed from the optical path.

A thermocouple was used to measure the temperature at the sample location. The temperature was gradually increased from ambient 23 °C to 260 °C at a rate of 60 °C/h in incremental steps. The transmission spectrum was acquired at each step once the temperature reached equilibrium. Then, the temperature was increased to the next step. Since the temperatures recorded at the oven controller and at the thermocouple were different, the temperature steps appeared uneven. The reported temperatures were those measured at the thermocouple, which has a greater precision (0.1 °C) than the oven controller. A final spectrum was acquired after the sample was left for 12 h in the oven at 260 °C.

2.2. Measurement of the Refractive Index Variation with Temperature and Exposure

Holographic optical elements are recorded by interference of two mutually coherent beams within the Bayfol® HX200 photopolymer material. As a result of the interference, a refractive index modulation is created within the active region of the material [5,6]. The refractive index spatial modulation $n_g(x)$, within the active region of the grating is approximated to have a sinusoidal structure and it is defined by Equation (1),

$$n_g(x) = n_o + \Delta n_g \cos\left(\frac{2\pi x}{\Lambda}\right) \tag{1}$$

where n_o is the average refractive index of the cured Bayfol® HX200 photopolymer, Δn_g is the refractive index modulation created by the interference between the signal and reference beams, and Λ is the period of the refractive index modulation.

Although both the average refractive index and refractive index modulation are assumed to be fixed after recording and processing, it is expected that they might change as a result of heat or refrigeration.

The refractive index measurement was performed using a Metricon 2010 Model (Pennington, NJ, USA). This system consists of a red laser at 637 nm, high refractive index prism, pneumatically operated coupling head, and photodetector. The prism and coupling head are heated to temperature values comprised between 24 °C and 100 °C with increments of 5 °C or less.

In Figure 1a, images of the fresh unexposed Bayfol® HX200 material are shown; in Figure 1b, pictures of the material after being fully cured by UV exposure are presented. The sample size was about 5×10 cm^2. The purple coloration of the fresh unexposed samples presented in Figure 1a was due to the photoinitiator that was responsible for the polymerization of the monomer under light exposure. The Bayfol® HX200 material uses a two-component system composed of a dye and an organoborate salt [26]. Once the material has been cured under UV light, the photoinitiator dye is left in a bleached state that is transparent to the visible light. Consequently, the cured samples presented in Figure 1b are colorless. Measurements on the unexposed material were performed with the room light off to avoid any photochemical reaction.

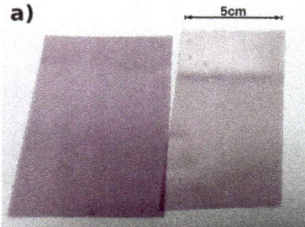 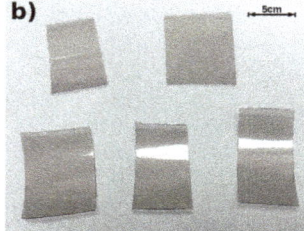

Figure 1. Pictures of Bayfol® HX200 film samples for (**a**) unrecorded/unexposed and (**b**) cured samples.

The Bayfol® HX200 photopolymer film was placed on the base of the high refractive index prism and the coupling head was brought pneumatically to ensure that the film was in near contact with the prism. Due to the soft nature of the Bayfol® HX200 photopolymer material, and the fact that any excessive pressure could alter the bonding of its chemical composition, the pneumatic air pressure was set to 20 psi. A piece of electrical tape was placed over the coupling head to minimize stress in the sample under test. For each measurement, the thermocouple controller was set to the desired temperature, and allowed enough time to heat up the sample prior to taking measurement.

For each refractive index versus temperature measurement, ten measurements were collected from different locations within a piece of the film. The mean of the data was then recorded as the average refractive index, n_0. The temperature of the prism and the coupling were changed to various values ranging between 24° and 100 °C with increments of 5 °C or less, and a linear function was fitted for all the collected refractive indices.

Extrapolation of the fitted function to an input temperature of 0 °C yields the average refractive index at 0 °C. The slope of the line indicates the variation of refractive index with temperature, $\frac{\partial n}{\partial T}$. The linear fit function can be expressed as

$$n_0 = \frac{\partial n}{\partial T} T + n[0]. \tag{2}$$

Since the holographic recording is usually done at room temperature, the refractive index modulation is calculated as the difference between the values of n_0 for unexposed/unrecorded film and cured film. The cured sample was laminated over 3 mm thick BK7 substrate glass (Thorlabs Inc., Newton, NJ, USA) and placed under 275W Xenon lamp (Osram HLX 64656-FNT

Xenophot, Osram Sylvania, Wilmington, MA, USA) for fourteen hours, followed by one hour exposure under the sun to make sure the sample was fully cured.

2.3. Linear Coefficient of Thermal Expansion

For volume phase holographic gratings such as the one recorded in Bayfol® HX200, the energy distribution depending on the angle and wavelength can be calculated using electromagnetic propagation theory or coupled wave analysis [5]. However, the direction (θ_B) and wavelength (λ_B) of the maximum efficiency can be predicted simply by using Bragg's equation:

$$sin(\theta_B) = \frac{\lambda_B}{2n\Lambda} \tag{3}$$

where n is the refractive index of the material and Λ is the distance between the modulation planes (Bragg planes).

When the material in which the hologram is recorded shrinks or swells, the spacing and angle of the modulation planes are affected, changing the $\vec{K}$ vector of the grating (with $|K| = 2\pi/\Lambda$). This impacts the diffraction angle and the diffracted spectrum of the hologram.

In the simplified configuration of a transmission hologram with no slant angle, only the lateral spacing of the modulation planes changes ($|K'| = 2\pi/(\Lambda + \Delta\Lambda)$), which influences the diffraction angle for a specific wavelength:

$$sin(\theta'_B) = sin(\theta_B)\frac{1}{1 + \frac{\Delta\Lambda}{\Lambda_B}} \tag{4}$$

Usually, this effect is not particularly visible since the material is laterally constrained by the substrate on which it is laminated. In the case where the two materials (hologram and substrate) have different coefficients of thermal expansion, the system is bending, introducing even more aberrations that could not easily be separated.

On the other hand, for a reflection grating with its modulation planes parallel to the substrate, a change of the material in the thickness direction affects the diffracted wavelength, keeping the diffraction angle the same:

$$\lambda'_B = \lambda_B \left(1 + \frac{\Delta\Lambda}{\Lambda_B}\right) \tag{5}$$

This wavelength change can easily be detected by a spectrometer and is not coupled to any other effect, making it a good candidate for the measurement of the material CTE.

In the more general case where the modulation planes have a slant angle different from 0 or $\pi/2$, a more rigorous calculation, such as coupled wave analysis, is required to determine the perturbation on both diffraction angle and wavelength.

It should also be noted that the wavelength shift in reflection hologram due to the material swelling has been extensively studied in other materials such as silver halide or dichromated gelatin [31]. This effect has been used for tuning the hologram color and for the production of three colors (red, green, and blue) 3D images with a single laser [32]. However, in the case of these collagen-based materials, the swelling is due to the absorption of water rather than because of CTE, and the effect is orders of magnitude larger than what is expected with temperature change.

For testing the temperature dependence on the spectrum of a reflection hologram, the Bayfol® HX200 sample was prepared as in the previous section: a 3×6 cm^2 sample was laminated on float glass, then encapsulated after the hologram was recorded. The reflection hologram was recorded with a 532 nm laser with one beam orthogonal to the sample and the other beam incident at 5° angle. This angle was introduced so the diffracted beam could be easily separated from the front face reflection.

The transmission spectrum (zero-order) was measured in a double-pass experiment where the illumination light was reflected back by a mirror located behind the sample. The orientation of the incident illumination and the mirror angle were optimized to achieve maximum diffraction efficiency at 532 nm. In Figure 2, the interpolation of the spectrum measured at room temperature with coupled wave analysis is presented. The best fit parameters were $\Lambda = 5637$ lp/mm (532 nm wavelength with 5° and 180° (=0° opposite direction) incidence angles), 7.2 µm effective thickness, and a refractive index modulation of $\Delta n = 0.026$.

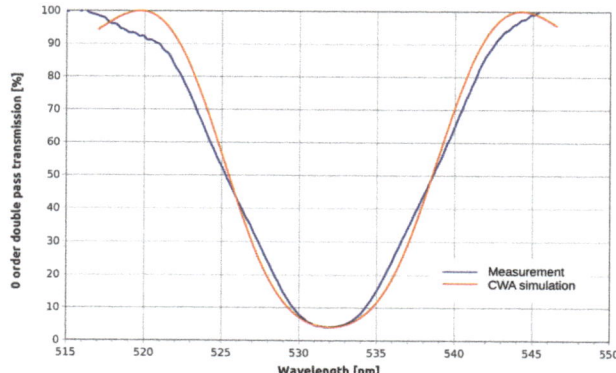

Figure 2. Double-pass transmission spectrum (zero-order) of the reflection hologram interpolated by coupled wave analysis.

The temperature was gradually raised from ambient to 170 °C over the course of 3 h, during which spectra were acquired at regular intervals. The sample was then allowed to cool down back to ambient temperature over 12 h. Finally, another temperature cycle was run with more spectra acquisition.

3. Results and Discussion

3.1. Thermal Degradation

The thermogravimetric analysis of both the Bayfol® HX200 photopolymer material and its cellulose triacetate backing (without photopolymer) is presented in Figure 3.

The photopolymer followed a multistage decomposition that started at 160 °C (first onset) and reached its midpoint at 180 °C. The backing film showed an initial degradation at the same temperature that is believed to be due to some photopolymer residue on our sample. A much larger weight loss appeared at 325 °C, which is consistent with the literature on cellulose triacetate [33].

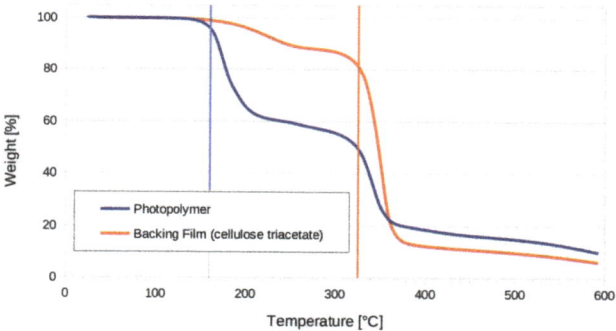

Figure 3. Thermogravimetric analysis of the Bayfol® HX200 photopolymer and the cellulose triacetate backing.

3.2. Spectral Measurement

In Figure 4, the results of the spectral measurement depending on temperature are presented. The initial spectrum taken at 47 °C shows the residual absorption of the sample composed of glass, Bayfol® HX200, and NOA60 glue, as well as the Fresnel reflection from the different interfaces. It can be noted from the picture taken of the sample before thermal treatment (Figure 5a) that the Bayfol® HX200 material is slightly yellow. No spectral change was observed up to 138 °C. The next spectrum recorded at 190 °C shows a decrease in transmission around 530 nm consistent with the thermogravimetric measurement. This indicates that the decomposition temperature was reached. As the temperature kept increasing, the transmission kept decreasing over the entire 400 nm to 840 nm band.

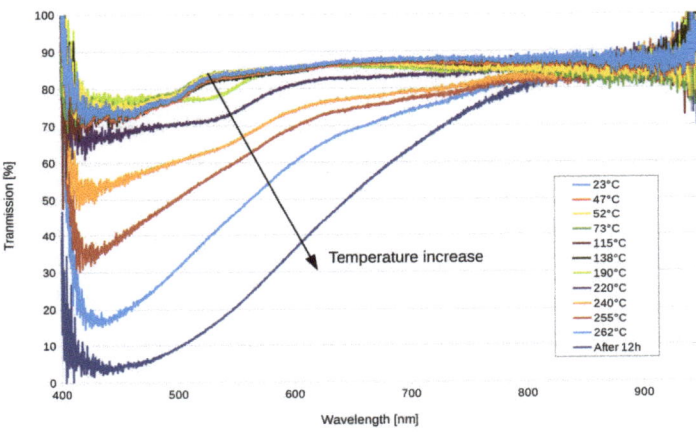

Figure 4. Transmission spectrum of the Bayfol® HX200 sample without hologram, depending on temperature.

Pictures of the sample before and after thermal treatment are presented in Figure 5. It can be noticed from Figure 5b that the brown coloration of the sample was only due to the Bayfol® HX200 material, since there was no color change for the NOA61 optical glue around the material. The optical transmittance of the sample before thermal treatment was 82% integrated over the entire visible spectrum (450–750 nm) and included the Fresnel reflection. After the thermal treatment (262 °C for 12 h), the optical transmittance was reduced to 27%.

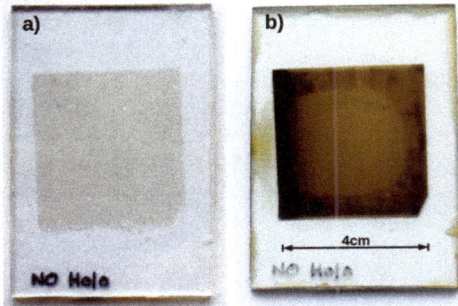

Figure 5. Pictures of the sample before thermal treatment (**a**), and after 12 h at 262 °C (**b**).

The precise chemical reaction responsible for the darkening of the photopolymer with temperature is beyond the scope of this research, but it can be deduced from similar behaviors observed in other polymers that the material undergoes thermal oxidation and further breakdown of the polymer chains [34]. To avoid these irreversible reactions, the Bayfol® HX200 material should be kept below 160 °C at all times.

3.3. Refractive Index

Figure 6 shows the change of the refractive index of the Bayfol® HX200 samples as they were heated to temperatures between 24 °C and 100 °C. The plots show that increased temperatures caused the refractive index to drop for both samples. The highest change is seen with the unrecorded samples as $\frac{\partial n}{\partial T}$ is 4.5×10^{-4} K^{-1}, when the cured samples have the value 3×10^{-4} K^{-1}.

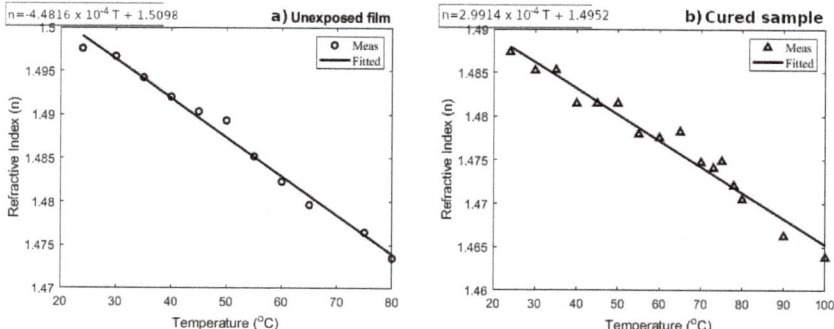

Figure 6. Refractive index variation depending on temperature for the (**a**) unexposed/unrecorded samples and (**b**) cured sample.

From the values of the refractive index presented in Figure 6, it can be noted that the refractive index change due to material curing by light exposure was around 0.015 for the cured sample. This Δn did not reach the maximum of 0.03 as indicated by the Bayfol® HX200 manufacturer data sheet [35]. Knowing the fact that the photopolymer materials rely on the photoinitiated cross-linking, monomer diffusion, and further polymerization to achieve high refractive index modulation [4], it can be understood that uniform illumination as used in this study does not allow for maximum refractive index change. In the case of large area illumination, the monomers could not diffuse far enough from the unexposed regions of the material to increase the material density and refractive index.

3.4. Linear Coefficient of Thermal Expansion

Figure 7 shows how the transmission spectrum of the reflection hologram drifted with temperature. It can be seen that from 23 °C to 171 °C, the transmission minimum drifted from 565 nm to 602 nm.

In Figure 8, the wavelength of the transmission minimum was plotted depending on temperature for two temperature cycles (back and forth) between 23 °C and 170 °C. During the initial temperature increase, an irreversible change occurred at 170 °C with a sudden increase in the minimum wavelength. After observation of the sample, we noticed that the NOA glue has softened and allowed the sample to shift from the cover glass. This shift induced a color change in the reflection spectrum of the hologram that can be seen in Figure 9b.

Figure 9 shows pictures of the reflection hologram sample before thermal treatment (left) and after the initial temperature increase to 170 °C (right). One can see some permanent change in the form of nonuniform color (pink or yellow) of the hologram after the thermal treatment. Note that the coloration of the sample was not due to material absorption, but to the diffraction from the hologram. It has to be noted that after this initial irreversible change, further temperature variation from 20 °C to 170 °C induces a linear and fully reversible change in diffracted wavelength.

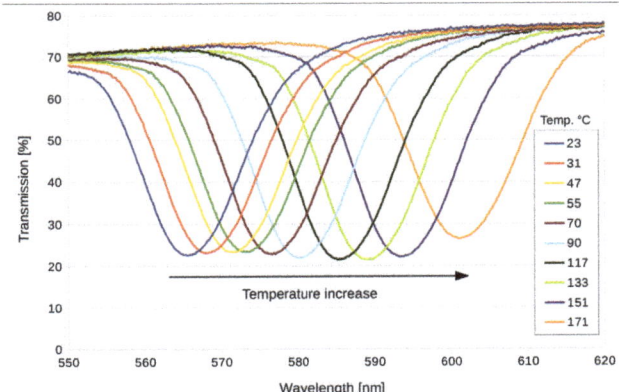

Figure 7. Transmission spectrum of the Bayfol® HX200 reflection hologram depending on temperature.

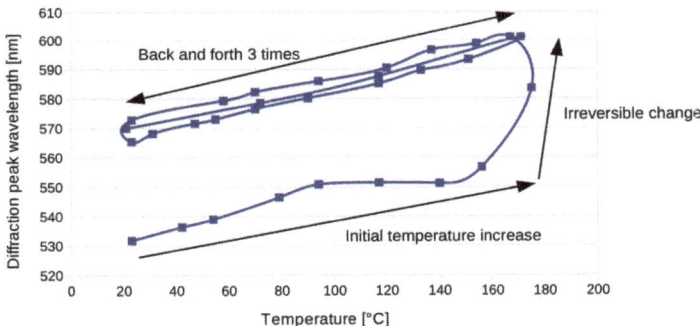

Figure 8. Wavelength minimum of the transmission spectra of the Bayfol® HX200 reflection hologram depending on temperature for two temperature cycles.

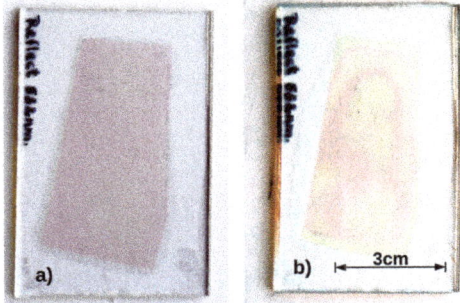

Figure 9. Pictures of the reflection hologram sample before thermal treatment (**a**) and after initial temperature increase from ambient temperature to 170 °C (**b**). Note that the pink coloration of the sample was not due to the material absorption but to the diffraction of the green wavelength that was no more present in the transmission spectra (white light − green = pink).

To calculate the linear coefficient of thermal expansion (CTE) for the Bayfol® HX200 material, we used Equation (5) to retrieve the spacing of the Bragg planes spacing from the measurement of the diffraction peak minimum wavelength as a function of temperature. This spacing parameter is presented in Figure 10 where it is interpolated by a linear regression. The origin of the straight line interpolation gives the original distance between the planes (L), and the slope gives the distance increase depending on temperature ($\Delta L/\Delta T$). The CTE is defined as:

$$\alpha = \frac{\Delta L}{L.\Delta T} = 384 \times 10^{-6} \ K^{-1} \qquad (6)$$

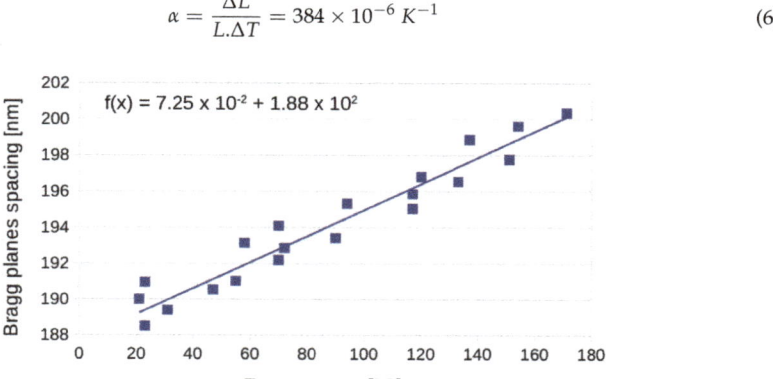

Figure 10. Bragg planes spacing for the Bayfol® HX200 reflection hologram depending on temperature for 1.5 temperature cycles.

The CTE calculated value $384 \times 10^{-6} \ K^{-1}$ for the Bayfol® HX200 is of the same order of magnitude as the literature values for other polymeric materials such as polypropylene $150 \times 10^{-6} \ K^{-1}$, polystyrene $150 \times 10^{-6} \ K^{-1}$, or teflon $175 \times 10^{-6} \ K^{-1}$ [36–38].

4. Conclusions

Thermogravimetric measurements show that the Bayfol® HX200 material does not suffer from any thermal degradation before 160 °C. Up to that temperature, holograms recorded in this material experience only reversible change, expressed as thermal dilatation between the Bragg planes. Irreversible change in the hologram diffraction spectrum started at 170 °C, and irreversible change in the transmission spectrum of the material was recorded starting at 190 °C.

Refractive index variation depending on the temperature was measured as $4.5 \times 10^{-4} \ K^{-1}$ for the unexposed material and $3.3 \times 10^{-4} \ K^{-1}$ once the material was fully cured. These changes are linear depending on the temperature over a range from 24 °C to 100 °C.

A CTE of $384 \times 10^{-6} \ K^{-1}$ was calculated by measuring the spectral shift of a reflection hologram depending on temperature. This color shift was due to the dilatation between the Bragg planes experienced when the photopolymer material swelled under temperature increase.

These results will be valuable to predict the characteristics of holographic optical elements and other holograms in applications and processes where thermal variation is expected.

Author Contributions: Conceptualization, P.-A.B., A.H.M., and E.B.; methodology, P.-A.B., A.H.M. and E.B.; manuscript writing, review, and editing, P.-A.B., A.H.M. and E.B. All authors have read and agreed to the published version of the manuscript.

Funding: Authors acknowledge the support of the NavAir under grant #N162-103-0420 NISE space section 219 program.

Conflicts of Interest: The authors declare no conflict of interest.

References

1. Jackin, B.J.; Jorissen, L.; Oi, R.; Wu, J.Y.; Wakunami, K.; Okui, M.; Ichihashi, Y.; Bekaert, P.; Huang, Y.P.; Yamamoto, K. Digitally designed holographic optical element for light field displays. *Opt. Lett.* **2018**, *43*, 3738–3741. [CrossRef]
2. Wu, H.Y.; Shin, C.W.; Kim, N. Full-color holographic optical elements for augmented reality display. In *Holographic Materials and Applications*; IntechOpen: London, UK, 2019.
3. Ferrara, M.A.; Striano, V.; Coppola, G. Volume holographic optical elements as solar concentrators: An Overview. *Appl. Sci.* **2019**, *9*, 193. [CrossRef]
4. Blanche, P.A. Holographic recording media and devices. In *Optical Holography—Materials, Theory and Applications*; Elsevier: St. Louis, MO, USA, 2020.
5. Kogelnik, H. Coupled wave theory for thick hologram gratings. *Bell Sysgtem Tech. J.* **1969**, *48*, 2909–2947. [CrossRef]
6. Mahamat, A.H.; Narducci, F.A.; Schwiegerling, J. Design and optimization of a volume-phase holographic grating for simultaneous use with red, green, and blue light using unpolarized light. *Appl. Opt.* **2016**, *55*, 1618–1624. [CrossRef]
7. Blanche, P.A. Introduction to Holography. In *Optical Holography–Materials, Theory and Applications*; Elsevier: St. Louis, MO, USA, 2020.
8. Behrmann, G.P.; Bowen, J.P. Influence of temperature on diffractive lens performance. *Appl. Opt.* **1993**, *32*, 2483–2489. [CrossRef]
9. Volodin, B.; Eichenberger, J.; Sayano, K. High-resolution compact imaging holographic Bragg grating spectrometer. In *Technical Digest. Summaries of Papers Presented at the Conference on Lasers and Electro-Optics*, Conference ed.; 1998 Technical Digest Series, (IEEE Cat. No. 98CH36178); IEEE: Washington. DC, USA, 1998; Volume 6, pp. 401–402.
10. Rakuljic, G.A.; Leyva, V. Volume holographic narrow-band optical filter. *Opt. Lett.* **1993**, *18*, 459–461. [CrossRef]
11. Steckman, G.J.; Liu, W.; Platz, R.; Schroeder, D.; Moser, C.; Havermeyer, F. Volume holographic grating wavelength stabilized laser diodes. *IEEE J. Sel. Top. Quantum Electron.* **2007**, *13*, 672–678. [CrossRef]
12. Bruder, F.K.; Fäcke, T.; Hagen, R.; Hansen, S.; Manecke, C.; Orselli, E.; Rewitz, C.; Rölle, T.; Walze, G. Thin combiner optics utilizing volume holographic optical elements (vHOEs) using Bayfol® HX200 photopolymer film. In *Digital Optical Technologies 2017*; International Society for Optics and Photonics: Bellingham, WA, USA, 2017; Volume 10335, p. 103350D.
13. Kress, B.C. Optical waveguide combiners for AR headsets: features and limitations. In *Digital Optical Technologies 2019*; International Society for Optics and Photonics: Bellingham, WA, USA, 2019; Volume 11062, p. 110620J.
14. Draper, C.T.; Bigler, C.M.; Mann, M.S.; Sarma, K.; Blanche, P.A. Holographic waveguide head-up display with 2-D pupil expansion and longitudinal image magnification. *Appl. Opt.* **2019**, *58*, A251–A257. [CrossRef]
15. Bruder, F.K.; Hansen, S.; Kleinschmidt, T.; Künzel, R.; Manecke, C.; Orselli, E.; Rewitz, C.; Rölle, T. Integration of volume holographic optical elements (vHOE) made with Bayfol® HX into plastic optical parts. In *Practical Holography XXXIII: Displays, Materials, and Applications*; International Society for Optics and Photonics: Bellingham, WA, USA, 2019; Volume 10944, p. 1094402.
16. Bruder, F.K.; Frank, J.; Hansen, S.; Künzel, R.; Manecke, C.; Meisenheimer, R.; Mills, J.; Pitzer, L.; Rölle, T.; Wewer, B. Expanding possibilities how to apply Bayfol® HX film into recording stacks and optical parts. In *Practical Holography XXXIV: Displays, Materials, and Applications*; International Society for Optics and Photonics: Bellingham, WA, USA, 2020; Volume 11306, p. 113060C.
17. Naik, G.M.; Mathur, A.; Pappu, S.V. Dichromated gelatin holograms: An investigation of their environmental stability. *Appl. Opt.* **1990**, *29*, 5292–5297. [CrossRef]
18. Mikulchyk, T.; Walshe, J.; Cody, D.; Martin, S.; Naydenova, I. Humidity and temperature response of photopolymer-based holographic gratings. In *Holography: Advances and Modern Trends IV*; International Society for Optics and Photonics: Bellingham, WA, USA, 2015; Volume 9508, p. 950809.
19. Naydenova, I.; Jallapuram, R.; Toal, V.; Martin, S. Hologram-based humidity indicator for domestic and packaging applications. In *Nanosensors, Microsensors, and Biosensors and Systems 2007*; International Society for Optics and Photonics: Bellingham, WA, USA, 2007; Volume 6528, p. 652811.
20. Naydenova, I.; Jallapuram, R.; Toal, V.; Martin, S. Characterisation of the humidity and temperature responses of a reflection hologram recorded in acrylamide-based photopolymer. *Sens. Actuators B Chem.* **2009**, *139*, 35–38. [CrossRef]

21. Liu, Y.; Li, W.; Ding, Q.; Yan, Z. Wavelength properties of DCG holograms under the conditions of different temperature and humidity. In *International Symposium on Optoelectronic Technology and Application 2014: Laser Materials Processing; and Micro/Nano Technologies*; International Society for Optics and Photonics: Bellingham, WA, USA, 2014; Volume 9295, p. 929514.
22. Lin, S.H.; Hsu, K.Y. Temperature effect in PQ:PMMA photopolymer. In *Photorefractive Fiber and Crystal Devices: Materials, Optical Properties, and Applications VI*; International Society for Optics and Photonics: Bellingham, WA, USA, 2000; Volume 4110, pp. 77–83.
23. Denning, R.G.; Blanford, C.F.; Urban, H.; Bharaj, H.; Sharp, D.N.; Turberfield, A.J. The Control of Shrinkage and Thermal Instability in SU-8 Photoresists for Holographic Lithography. *Adv. Funct. Mater.* **2011**, *21*, 1593–1601. [CrossRef]
24. Bigler, C.M.; Blanche, P.A.; Sarma, K. Holographic waveguide heads-up display for longitudinal image magnification and pupil expansion. *Appl. Opt.* **2018**, *57*, 2007–2013. [CrossRef]
25. Vorndran, S.D.; Chrysler, B.; Wheelwright, B.; Angel, R.; Holman, Z.; Kostuk, R. Off-axis holographic lens spectrum-splitting photovoltaic system for direct and diffuse solar energy conversion. *Appl. Opt.* **2016**, *55*, 7522–7529. [CrossRef]
26. Bruder, F.K.; Fäcke, T.; Rölle, T. The chemistry and physics of Bayfol® HX film holographic photopolymer. *Polymers* **2017**, *9*, 472. [CrossRef]
27. Bruder, F.K.; Fäcke, T.; Grote, F.; Hagen, R.; Hönel, D.; Koch, E.; Rewitz, C.; Walze, G.; Wewer, B. Mass production of volume holographic optical elements (vHOEs) using Bayfol (R) HX photopolymer film in a roll-to-roll copy process. In *Practical Holography XXXI: Materials and Applications*; International Society for Optics and Photonics: Bellingham, WA, USA, 2017; Volume 10127, p. 101270A.
28. Bruder, F.K.; Hansen, S.; Manecke, C.; Orselli, E.; Rewitz, C.; Rölle, T.; Wewer, B. Wavelength multiplexing recording of vHOEs in Bayfol HX photopolymer film. In *Digital Optics for Immersive Displays*; International Society for Optics and Photonics: Bellingham, WA, USA, 2018; Volume 10676, p. 106760H.
29. Covestro. *Bayfol® HX200 Description and Application Information*; Technical Report; Bayer 04 Leverkusen: Leverkusen, Germany, 2018.
30. Bruder, F.K.; Hansen, S.; Kleinschmidt, T.; Künzel, R.; Manecke, C.; Orselli, E.; Rewitz, C.; Rölle, T. How to integrate volume holographic optical elements (vHOE) made with Bayfol HX film into plastic optical parts. In *Holography: Advances and Modern Trends VI*; International Society for Optics and Photonics: Bellingham, WA, USA, 2019; Volume 11030, p. 110300C.
31. Bjelkhagen, H.; Brotherton-Ratcliffe, D. *Ultra-Realistic Imaging: Advanced Techniques in Analogue and Digital Colour Holography*; CRC Press: Boca Raton, FL, USA, 2013.
32. McGrew, S.P. Color control in dichromated gelatin reflection holograms. In *Recent Advances in Holography*; International Society for Optics and Photonics: Bellingham, WA, USA, 1980; Volume 215, pp. 24–31.
33. Li, X.G. High-resolution thermogravimetry of cellulose esters. *J. Appl. Polym. Sci.* **1999**, *71*, 573–578. [CrossRef]
34. Visakh, P.; Nazarenko, O.B. Thermal degradation of polymer blends, composites and nanocomposites. In *Thermal Degradation of Polymer Blends, Composites and Nanocomposites*; Springer: Cham, Switzerland, 2015; pp. 1–16.
35. *Bayfol HX200 Technical Data Sheet*; Covestro: Leverkusen, Germany, 2020.
36. Jayanna, H.; Subramanyam, S. Thermal expansion of irradiated polypropylene from 10–340 K. *J. Mater. Sci.* **1993**, *28*, 2423–2427. [CrossRef]
37. Pye, J.E.; Roth, C.B. Above, below, and in-between the two glass transitions of ultrathin free-standing polystyrene films: Thermal expansion coefficient and physical aging. *J. Polym. Sci. Part Polym. Phys.* **2015**, *53*, 64–75. [CrossRef]
38. Kirby, R.K. Thermal expansion of polytetrafluoroethylene (Teflon) from −190 to +300 °C. *J. Res. Natl. Bur. Stand.* **1956**, *57*, 91–94. [CrossRef]

Publisher's Note: MDPI stays neutral with regard to jurisdictional claims in published maps and institutional affiliations.

© 2020 by the authors. Licensee MDPI, Basel, Switzerland. This article is an open access article distributed under the terms and conditions of the Creative Commons Attribution (CC BY) license (http://creativecommons.org/licenses/by/4.0/).

Article

Enhanced Magneto-Optic Properties in Sputtered Bi- Containing Ferrite Garnet Thin Films Fabricated Using Oxygen Plasma Treatment and Metal Oxide Protective Layers

V. A. Kotov [1], M. Nur-E-Alam [2,*], M. Vasiliev [2], K. Alameh [2], D. E. Balabanov [3] and V. I. Burkov [3]

[1] Institute of Radio Engineering and Electronics, Russian Academy of Sciences, 11 Mohovaya St, Moscow 125009, Russia; kotov.slava@gmail.com
[2] Electron Science Research Institute, Edith Cowan University, Joondalup, WA 6027, Australia; vasiliev.mikhail@gmail.com (M.V.); k.alameh@ecu.edu.au (K.A.)
[3] Moscow Institute of Physics and Technology, 9 Institutski Per., Dolgoprudny 141700, Russia; dima-mipt@mail.ru (D.E.B.); optikcentr@mail.mipt.ru (V.I.B.)
* Correspondence: m.nur-e-alam@ecu.edu.au

Received: 30 September 2020; Accepted: 10 November 2020; Published: 12 November 2020

Abstract: Magneto-optic (MO) imaging and sensing are at present the most developed practical applications of thin-film MO garnet materials. However, in order to improve sensitivity for a range of established and forward-looking applications, the technology and component-related advances are still necessary. These improvements are expected to originate from new material system development. We propose a set of technological modifications for the RF-magnetron sputtering deposition and crystallization annealing of magneto-optic bismuth-substituted iron-garnet films and investigate the improved material properties. Results show that standard crystallization annealing for the as-deposited ultrathin (sputtered 10 nm thick, amorphous phase) films resulted in more than a factor of two loss in the magneto-optical activity of the films in the visible spectral region, compared to the liquid-phase grown epitaxial films. Results also show that an additional 10 nm-thick metal-oxide (Bi_2O_3) protective layer above the amorphous film results in ~2.7 times increase in the magneto-optical quality of crystallized iron-garnet films. On the other hand, the effects of post-deposition oxygen (O_2) plasma treatment on the magneto-optical (MO) properties of Bismuth substituted iron garnet thin film materials are investigated. Results show that in the visible part of the electromagnetic spectrum (at 532 nm), the O_2 treated (up to 3 min) garnet films retain higher specific Faraday rotation and figures of merit compared to non-treated garnet films.

Keywords: magneto-optics; mcd; faraday rotation; figure of merit; polarization; oxygen plasma treatment

1. Introduction

Since several decades ago, magneto-optic (MO) applications of garnet materials were well-known. Bismuth (Bi)-substituted garnet materials for various MO applications attract the attention of researchers in this field, aimed at developing innovative high-performance garnet materials or finding ways of improving their properties. Also, from the practical point of view, MO garnet materials of these composition types with high-performance are relevant to the context of manufacturing of next-generation ultra-fast optoelectronic devices, such as light intensity switches and modulators, high-speed flat panel displays and high-sensitivity sensors [1–14]. Therefore, it is important, nowadays, to obtain MO materials of optimized material composition stoichiometry possessing a high figure of merit and a low coercive field. This can typically be achieved with garnets having high Bi-substitution levels. In recent years, significant research activities have been reported in the field of the manufacture and characterization of new-generation ultrathin films of yttrium iron garnet (YIG) and related materials, of thicknesses ranging from several nanometers to several tens of nanometers. The record-low optical losses of YIGs in the ultrahigh frequency (UHF) spectral region make them attractive for the development of spintronics and modern microwave devices. The low losses of YIGs are due to the small damping parameter $\alpha \approx 3 \times 10^{-5}$ (the ferromagnetic resonance linewidth of less than 0.5 Oe at 9 GHz). Thin films of ferromagnetic metals lag behind YIG materials in performance (in terms of the damping parameter characteristics) by two orders of magnitude, thus enabling a significant reduction in switching current for spin valve devices, which are the key components used to develop magnetic sensors, hard disk read heads and magnetic random access memories (MRAM) [1]. Another potential practical area that can benefit from the use of thin YIG films is related to the development of different nano-electronic device types, which utilize the phenomenon of spin current generation by magnetostatic spin waves propagating in thin YIG films possessing small damping parameters [5]. Bismuth-substituted iron garnets, being ferrimagnetic dielectrics possessing giant specific Faraday rotation across the visible and near-infrared spectral regions, also represent the most promising MO materials for use in different MO devices, such as magnetic photonic crystals (MPC), non-reciprocal MO elements, Faraday-effect ultrafast MO modulators, magnetic field-controlled multilayer MO waveguiding structures, hybrid multiferroics-based multilayers and other applications in photonics [15–17]. From the point of view of the practical applications of ferrite garnets in hybrid integrated-optics circuits, the most promising fabrication approach is RF magnetron sputtering of amorphous-phase garnet films onto substrates such as gadolinium gallium garnet (GGG), kept at room temperature (or between 100–400 °C), followed by annealing crystallization processes run at temperatures between 490–650 °C. The properties of a transitional layer forming between the substrate and deposited film are a defining factor, which governs the annealing crystallization process since the crystallization processes of these garnet film structures start from the transitional layer region. When using the thin or ultrathin ferrite garnet layers in bilayer-type structures involving garnet film as spin wave generator and a nanoscale platinum film as spin current detector, at the forefront is the problem of the uniformity of the magnetic properties in thin or ultrathin garnet films and whether these also possess record-low damping parameters or small ferromagnetic resonance (FMR) linewidths near 1 Oe. In addition to this, the application of additional or protective layers or films is very useful for the topographic mapping of the sensitive media [18–20]. There are literature reports, presenting the data showing that in thin and ultrathin iron-garnet films, there exist significant variations in both the composition and magnetic properties across the film thickness [5]. For example, near the substrate-film boundary region, in epitaxially-grown iron garnet films, dependent on growth conditions, complex transitional layers of thickness ranging between several nm up to 250 nm, may form [21]. It is important to note that in these films, the transitional-layer thickness (as evaluated using Curie temperature measurements), may also reach 250 nm. With increasing film thickness, up to 2 µm, a constant Curie temperature value is observed.

On the other hand, oxygen plasma treatment is an attractive and widely used technique on both the experimental and industrial scales to improve thin-film technology without introducing any complexity into the material stoichiometry [22–24]. Oxygen plasma treatment is a well-known method used to clean the substrates for the development of thin films, since the oxygen plasma treatment enhances the adhesion of thin films to the substrates. The optimized oxygen plasma treatment process can enhance the thin film's bonding strength (surface energy) and adequately modify the film surfaces for various microelectronics and optoelectronics devices without affecting the entire nanostructure of the devices [22–27]. J. W. Roh et al. have reported that the oxygen plasma- assisted wafer bonding process is very effective and crucial for the fabrication of integrated optical waveguide isolators. They treated the surfaces of GGG substrates by oxygen plasma for 30 s with a radio frequency (RF) plasma power of 100 W under oxygen pressure of 0.3 Torr and observed high bonding strength and hydrophilicity without any voids in the interface in bonding of Indium phosphide (InP) thin films to

Gd$_3$Ga$_5$O$_{12}$ (GGG) substrates [26]. K. H. Chen et al. have applied the oxygen plasma treatment process in low temperature environment to improve the electrical and physical properties of as-deposited (Ba$_{0.7}$Sr$_{0.3}$)(Ti$_{0.9}$Zr$_{0.1}$)O$_3$ (BSTZ) thin films [27]. They have reported that the oxygen plasma treatment affects the chemical bonding state and crystalline structure to help reduce the density of interface states, oxygen vacancies and defects for as-deposited BSTZ thin films and enhance the capacitance of the films. Growing high-quality thin films of various oxide and metal-oxide-based materials, including MO garnets, on various substrates, is typically accomplished by creating oxygen plasma and allowing extra oxygen input with argon (Ar), Nitrogen (N$_2$) or Hydrogen (H) or Helium (He) plasma during the deposition process [28–37]. However, to the best of our knowledge, using post-deposition oxygen plasma treatment on as-deposited highly Bi-substituted iron garnet thin films, prior to the annealing crystallization processes, has never been reported, at least not in conjunction with MO quality measurements.

To improve the properties of highly Bi-substituted metal doped iron garnet thin film materials, we propose two new and modified process sequences for annealing crystallization of garnet thin films. The new method for the manufacture of high-performance ultrathin garnet films is the provision of a thin (2–20 nm) protective bismuth oxide (Bi$_2$O$_3$) layer, which assists in the crystallization of the garnet layer. In this case, during the initial stage, a 20–60 nm amorphous-phase film of a nanocomposite co-sputtered material type of (Bi$_2$Dy$_1$Fe$_4$Ga$_1$O$_{12}$ + Bi$_2$O$_3$), is deposited onto a GGG or a glass substrate. The excess bismuth oxide content relative to the stoichiometric composition of ferrite garnet is kept between 10–40 vol %. The second stage involves the deposition (also by RF sputtering) of a protective bismuth-oxide layer of thickness between 2 nm–20 nm onto these amorphous nanocomposite films. The obtained two-layer structure is then subjected to annealing crystallization in an air atmosphere, at a temperature between 490 °C and 650 °C, for 1 h. Results show that the MO performance characteristics of the samples (nanocomposite material of composition type Bi$_2$Dy$_1$Fe$_4$Ga$_1$O$_{12}$ + Bi$_2$O$_3$ with a protective Bi$_2$O$_3$ layer) exceed, by more than a factor of two, the corresponding parameters obtained in identical material systems fabricated without this additional protective layer. We also report on the studies of the Faraday rotation and its dispersion (conducted in the 400 nm–600 nm interval), as well as the magnetic circular dichroism (MCD, performed in between 300 nm–600 nm). Results demonstrate a two-fold improvement in the MO characteristics of oxide-protected garnet films, due to both the increased bismuth substitution levels and the prevention of bismuth evaporation from the subsurface film regions.

Secondly, we apply oxygen plasma treatment on as-deposited garnet samples immediately after deposition and then follow the previously established (composition-dependent) high-temperature annealing processes to crystallize the garnet thin films. We investigate the effects of post-deposition oxygen plasma treatment on the MO properties of RF sputtered garnet thin-film layers, synthesized using two different types of Bi-substituted garnets. The oxygen plasma treated and non-treated garnet thin films are characterized and analyzed after running the annealing processes. In the conducted experiments, we repeatedly noticed that the post-deposition low- temperature oxygen plasma treatment improves their material properties, especially the Faraday rotation per unit film thickness and the optical absorption coefficients, thus leading to obtaining a higher MO figure of merit compared to that of non-treated annealed garnet layers.

2. Background and Transitional Layer Properties

When using a single solution-melt, depending on the epitaxial growth temperature and the supercooling magnitude, the transitional layer thickness can vary widely, from 5 nm to around 250 nm, with the transitional layer being possibly composed of several intermediate layers. For example, a growth regime with bismuth-containing solution-melt supercooling near $\Delta T \sim 150$ °C at a growth temperature around 750 °C leads to the appearance of an intermediate transitional sublayer of thickness around 100 nm at the substrate-film boundary (Curie temperature of the iron garnet composition being 225 °C). As a result, in this thickness interval, the epitaxial growth process occurs under the conditions that the growth rate is being limited by the crystallization rate at the substrate-film boundary. Past this stage, a thick transitional sublayer appears, of thickness near 150 nm, within which the Curie temperature reduces from 225 °C to 215 °C. At the same time, for this material, the effective field of the uniaxial magnetic anisotropy, defined as $H_k^{eff} = H_k - 4\pi M_s$, changes smoothly from H_k^{eff} = 1500 Oe at the epitaxial layer thickness h = 30 nm, to H_k^{eff} = 2100 Oe, at h = 250 nm. Studies of the lattice parameter dependency on the epitaxial layer thickness conducted in the thickness range between 250 nm and 1μm for Bi-substituted ferrite garnet films showed that both the Curie temperature of material and the lattice parameter do not change and are equal to T_C = 215 °C and a_f = 12.401 Å, respectively. Within the starting region of the transitional layer, the corresponding measured values were T_C = 225 °C and a_f = 12.412 Å, respectively [1]. According to the data reported in Reference [5], an increase in the Bi substitution by 1 formula unit (f.u.) within epitaxial films has led to an increase in the garnet lattice parameter by Δa_f = 0.0828 Å. Therefore, increasing the Bi-substitution from 0.3 f.u. to 1.43 f.u. should lead to an increase of the transverse lattice parameter from a_f = 12.401 Å to a_f = 12.412 Å and hence, the data on the T_c and lattice parameter near the film-substrate boundary do not match well [1,38].

At a growth temperature around 980 °C and solution melt supercooling of around ΔT = 5 °C, it is possible to grow an epitaxial garnet layer of thickness around several microns, within which the Curie temperature remains practically constant across the entire volume of the epitaxial layer. It is important to note that, when fabricating thin and ultrathin bismuth-substituted iron-garnet layers, the formation of transitional layers near the film-substrate boundary may also take place due to the partial amorphization of the substrate surface occurring during the pre-deposition argon-plasma bombardment as a result of additional substrate-cleaning measures. The Ar^+ ion energies may, in this case, reach between tens of eV to several keV. Another cause of the significant changes in the composition of film with thickness and the related changes in the magnetic properties of films, is the annealing crystallization procedure, which takes place within the (composition-dependent) temperature range from 490 °C to 650 °C [1]. Etching of GGG substrates undertaken prior to the epitaxial film growth leads, at best, to the root mean squre RMS surface roughness of the substrate surface being near ~0.25 nm. Usually, in Liquid Phase Epitaxy (LPE)-grown iron garnet films fabricated while keeping constant melt temperature during growth, a significant reduction of the Bi substitution content is observed across the film layer thickness, towards the direction of the film-air boundary. During the experiments aimed at finding the optimum temperature of epitaxial growth, it has been found that, at a growth temperature between 950 °C and 980 °C and melt supercooling near ΔT = 5–10 °C, it is possible to manufacture films with a constant Curie temperature, within ΔT_c = 3 °C [1]. In this study, several batches of Bi-containing thin-ferrite garnet-type films are fabricated and characterized in order to better understand the annealing crystallization processes for the synthesis Bi-substituted ferrite garnets (which initially are found to be in an amorphous phase after RF magnetron deposition).

3. Garnet Layers Sputter-Deposition and Annealing Process and Parameters

Multiple batches of single-layer bismuth-substituted garnet compounds doped with dysprosium and gallium and bi-layer structures (garnet layer covered by a top thin protective oxide layer) have been prepared on glass and monocrystalline garnet substrates using the RF magnetron sputtering technique. The sputtering targets used had nominal compositions of $Bi_2Dy_1Fe_4Ga_1O_{12}$, $Bi_{2.1}Dy_{0.9}Fe_{3.9}Ga_{1.1}O_{12}$, $Bi_{1.8}Lu_{1.2}Fe_{3.6}Ga_{1.4}O_{12}$ and Bi_2O_3. From our previous work, we had found that the films of co-sputtered composition type ($Bi_2Dy_1Fe_4Ga_1O_{12}$ sputtered with excess Bi_2O_3) possessed simultaneously a high Faraday rotation and the necessary level of uniaxial magnetic anisotropy to orient the magnetization of the films in the direction perpendicular to the film plane [5].

Some of the as-deposited garnet layers were treated with oxygen plasma exposure immediately after the deposition process before the high temperature crystallization process was performed. The process parameters used to prepare the garnet layers, including sputter deposition, oxygen plasma exposure and annealing crystallization, are detailed in Table 1. O_2 plasma treatment was conducted using YZD08-5C plasma cleaner (purchased through Alibaba.com) for 0.5–5 min. The plasma-treated and the non-treated samples were then annealed by using the optimized annealing regimes found in previous annealing experiments for this composition of garnet layers [3,4]. The film quality and the properties of annealed garnet samples were first characterized in terms of the specific Faraday rotation and MO figure of merit at 532 nm.

Table 1. Summary of process parameters used to prepare the garnet layers.

Sample Preparation Stage	Process Parameters	Values & Comments
Garnet layers deposition	Sputtering target stoichiometry oxide-mixed garnet targets	$Bi_2Dy_1Fe_4Ga_1O_{12}$, $Bi_{2.1}Dy_{0.9}Fe_{3.9}Ga_{1.1}O_{12}$, $Bi_{1.8}Lu_{1.2}Fe_{3.6}Ga_{1.4}O_{12}$ and Bi_2O_3
	Base pressure	4–5×10^{-6} Torr
	Argon (Ar) pressure	≈ 2 mTorr
	Substrate stage temperature	Room Temperature 21–23 °C
	Substrate stage rotation rate	16–17 rpm
Oxygen plasma treatment	Base pressure	750 mTorr
	Oxygen flow	0.2 sccm/min
	RF power densities	40 W
	Plasma exposure time	30 s to 5 min

In this work, the post-deposition annealing processes were run using a conventional temperature-controlled and heating-rate-controlled oven in the temperature range of 490 to 650 °C.

Several batches of simple (double and triple)-layer-type all-garnet heterostructures were also manufactured, investigated and their optimized process parameters and properties (i.e., optimizing the heterostructure annealing regimes and characterizing the crystallization behavior, inter-material compatibility and microstructural properties) were reported by our group. Our previously published data confirmed the annealing crystallization behavior of Bi-substituted iron garnet and garnet-oxide composites deposited onto various substrate types [3–5].

4. Results

4.1. Spectral Dependencies of Faraday Rotation Measured for Nanocrystalline Films of Composition $Bi_2Dy_1Fe_4Ga_1O_{12}$

Figure 1 shows the spectral dependency of the specific Faraday rotation, where a peak at 494 nm is observed, for a $Bi_2Dy_1Fe_4Ga_1O_{12}$ film of around 150 nm thickness. When measuring the MO characteristics in samples of less than 20 nm thickness, across the temperature interval between 8K–200K (−265.15 to −73.15 °C), we observed magnetic circular dichroism (MCD) spectra similar to these typical for nanocrystalline Bi-substituted ferrite garnets with thicknesses between 500–1000 nm. During this study, we measured the MCD spectra between 250–600 nm.

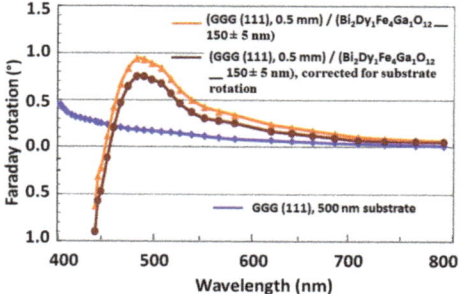

Figure 1. Spectral dependencies of Faraday rotation measured for nanocrystalline films of composition $Bi_2Dy_1Fe_4Ga_1O_{12}$ at the saturated magnetization state.

Y. Sun et al. has reported PLD-manufactured YIG films on GGG substrates with FMR linewidth of 3.4 Oe, defined as the interval between the extrema of the derivative of the FMR absorption line at 10 GHz. The film surface roughness determined by scanning probe microscopy was between 1–3 nm [39]. The years 2010 and 2011 witnessed the birth of a new paradigm in the discipline of spintronics—"spintronics using yttrium iron garnets [1,15–17,40,41]. The significance of this research field originates from two features of yttrium iron garnet ($Y_3Fe_5O_{12}$, YIG) materials: (1) extremely small damping factor and (2) electrically-insulating property.

4.2. Scpectral Dependencies of Magnetic Circular Dichroism for Standard Nanocomposite-Type Samples of $Bi_2Dy_1Fe_4Ga_1O_{12} + Bi_2O_3$

In this study, we measured the magnetic circular dichroism (MCD) spectra in the wavelength range of 250–600 nm in nanocomposite films of bismuth-containing ferrite garnets with an excess of bismuth oxide. The results of studying the spectral dependence of magnetic circular dichroism in a nanocomposite film of the $Bi_2Dy_1Fe_4Ga_1O_{12} + Bi_2O_3$ system in the spectral range from 250 to 600 nm are shown in Figure 2. The sign of the MCD effect is opposite to the sign of MCD observed in films of ferrite garnets of composition $(YBi)_3Fe_5O_{12}$ [40]. This is because the studied sample has a magnetic compensation point at a temperature above room temperature. Note that, for $(YBi)_3Fe_5O_{12}$ samples, the tetrahedral magnetic sublattice of a ferrite garnet is oriented along the applied magnetic field, however, by the substituting yttrium ions by dysprosium ions and iron ions by gallium ions in the dodecahedral ferrite garnet sublattice reverse the magnetization orientation.

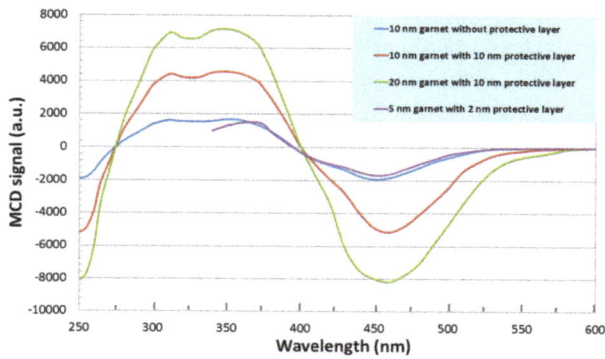

Figure 2. Spectral dependences of the magnetic circular dichroism of ferrite garnet films, in the spectral range from 250 to 600 nm.

4.3. $Bi_{2.1}Dy_{0.9}Fe_{3.9}Ga_{1.1}O_{12}$ and $Bi_{1.8}Lu_{1.2}Fe_{3.6}Al_{1.4}O_{12}$ Garnet Layers Annealed Followed by Post-Deposition Oxygen Plasma Treatment

X-ray diffraction (XRD) traces obtained for as-deposited and annealed $Bi_{2.1}Dy_{0.9}Fe_{3.9}Ga_{1.1}O_{12}$ films using a Siemens 5000D x-ray diffractometer are shown in Figure 3. The presence of a weak broad hump at ~38.2° in the XRD pattern of the as-deposited garnet samples is attributed to the amorphous phase of the samples just after deposition as well as to the post oxygen plasma exposure. However, following annealing at 580 °C and higher, the broadened hump at 38.2° is turned into a small but significantly noticeable, peak (512) together with a number of stronger peaks consistent with the primary garnet phase representing their crystallization stage and the nanocrystalline microstructure of the garnet films. All identified XRD peaks and their angular positions with the half maximum-line width (FWHM) values were determined using the Jade 9 (MDI Corp.) software package (Peak-listing option). The lattice constant and crystallite sizes for the synthesized garnet-type materials were calculated using the standard procedures followed in Ref [3]. It can be concluded that all the garnet layers present the crystalline phage however, from our experiments we observed that the Oxygen plasma treated samples were able to annealed at a comparatively low temperature compared to that of the non-treated garnet samples.

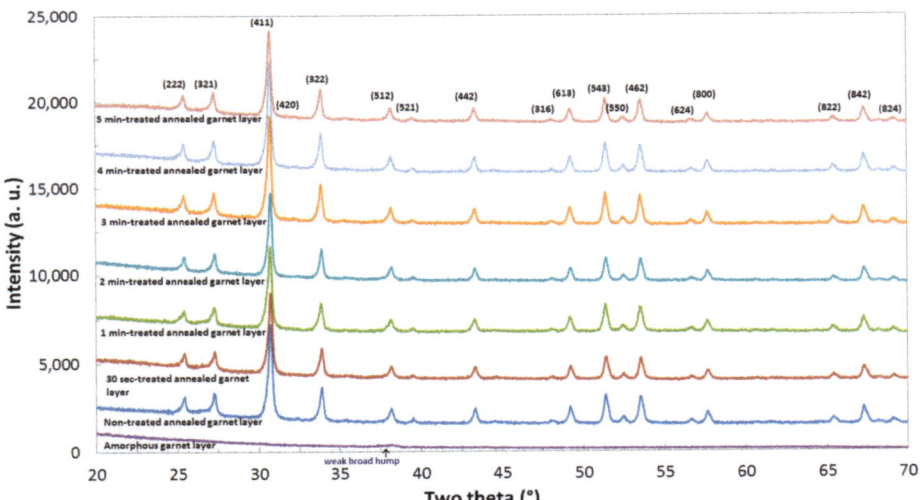

Figure 3. Measured X-ray diffraction (XRD) patterns in as-deposited and annealed $Bi_{2.1}Dy_{0.9}Fe_{3.9}Ga_{1.1}O_{12}$ garnet layers.

To investigate the effects of post-deposition oxygen plasma treatment on the optical and magneto-optical behaviours of $Bi_{2.1}Dy_{0.9}Fe_{3.9}Ga_{1.1}O_{12}$ type thin films, we derived the optical absorption coefficients, measured the specific Faraday rotation and calculated the MO figure of merit at a certain spectral wavelength (532 nm). The transmission spectra of all annealed garnet layers were spectrally fitted with the modeled transmission spectra to determine the optical absorption on the garnet layers by using the MPC software reported in Reference [42]. The plasma-treated samples showed slightly higher specific Faraday rotation at 532 nm compared to the non-treated garnet samples and this was repeatedly observed in all batches of plasma-treated annealed samples. However, significantly lower optical losses (optical absorption) were observed in the O_2-plasma-treated (up to 3 min) garnet layers than that of non-treated garnet films, thus leading to an improved MO figure of merit ($Q = 2 \times \Theta_F/\alpha$, where Θ_F is the specific Faraday rotation and α is the absorption coefficients) as shown in Figures 4 and 5.

Estimated errors in films' thicknesses (within ± 5% accuracy) as well as in Faraday rotation were accounted for during the calculations of the MO figures of merit.

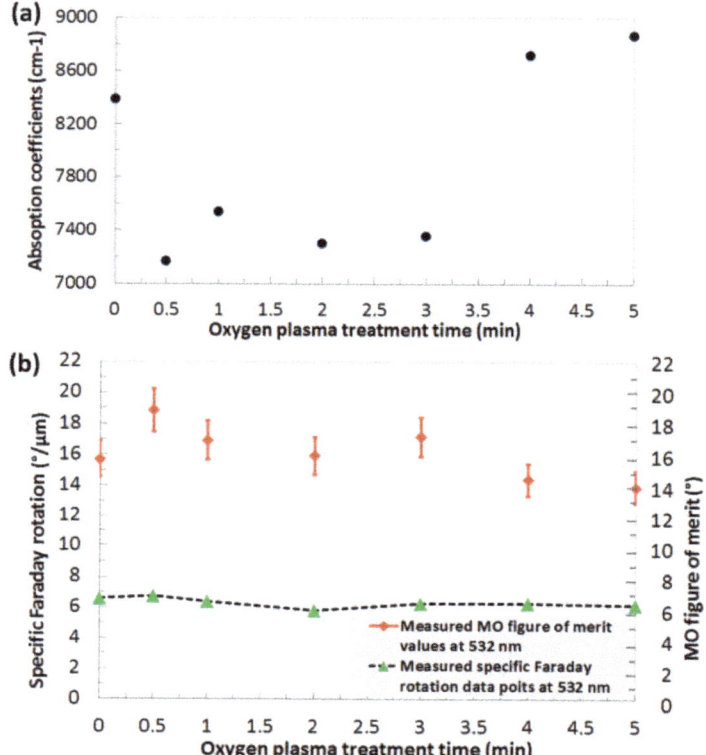

Figure 4. Measured optical absorption coefficients (**a**), specific Faraday rotation and magneto-optic (MO) figure of merit data points (**b**) at 532 nm in optimally annealed O_2 plasma treated and non-treated $Bi_{2.1}Dy_{0.9}Fe_{3.9}Ga_{1.1}O_{12}$ garnet layers.

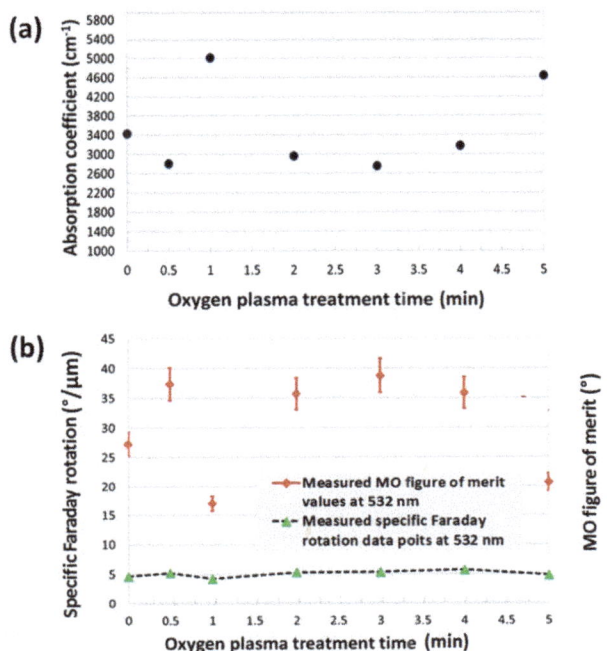

Figure 5. Measured optical absorption coefficients (**a**), specific Faraday rotation and MO figure of merit data points (**b**) at 532 nm in optimally annealed O_2 plasma treated and non-treated $Bi_{1.8}Lu_{1.2}Fe_{3.6}Al_{1.4}O_{12}$ garnet layers.

The best MO properties were obtained in the sample that was exposed to oxygen plasma for only 30 s inside the high vacuum chamber. This sample showed the highest specific Faraday rotation, lowest optical absorption at 532 nm. The highest MO figure of merit, about 19 degrees, was obtained in the 30-s plasma-treated garnet layer of composition type $Bi_{2.1}Dy_{0.9}Fe_{3.9}Ga_{1.1}O_{12}$ whilst the highest MO figure of merit, above 35 degrees, was obtained in plasma-treated garnet layers of composition type $Bi_{1.8}Lu_{1.2}Fe_{3.6}Al_{1.4}O_{12}$. It is important to note that the specific Faraday rotation slightly decreased with increasing the plasma treatment time, however, the absorption coefficient increased significantly with the plasma exposure time, and, consequently, reduced the MO quality of the garnet layers after 3 min.

The surface roughness profile of the garnet thin films (oxygen plasma treated and non-treated) were studied and it was found that all annealed samples exhibited smooth, uniform and consistent morphology across film surfaces except some minor micro-cracks (occurred due to substrates high-temperature expansion and also for the lattice mismatch of the garnet layer and the substrate). Garnet films exposed to oxygen plasma for 30 s possessed the best surface quality with RMS roughness value (Rq) of around 0.5 nm, as can be seen in Figure 6. This indicates that such films have significant surface effects while other samples including the non-treated garnet layer showed RMS value over 1 nm. From the above discussion, it can be concluded that the O_2 plasma treatment effectively interacts with the material surface layer and reduces the surface roughness without intruding into the entire layer structure. This changes the surface energy of the garnet layers and helps the garnet layers get more oxygen diffusion during the annealing crystallization process, leading to better MO properties compared to the non-treated garnet layers. Note that, films treated with O_2 plasma for longer than 30 s displayed surface larger nanoscale grain features.

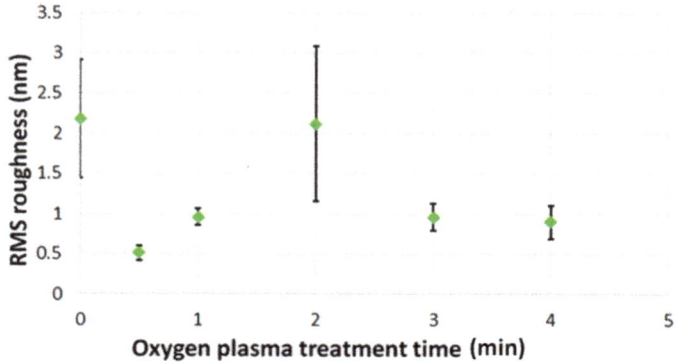

Figure 6. Measured root mean square RMS surface roughness values vs. oxygen plasma exposure time, for the various developed garnet thin films.

It was observed that the MO figure of merit decreased with the increase in plasma treatment time though the specific Faraday rotation remained higher compared to that of the non-treated annealed garnet layer. From the overall experimental datasets, we can conclude that the oxygen plasma treatment helps the amorphous garnet layers to crystallize with less effect added from foreign contaminants on the layer surface, compared to the non-treated samples. This is an ongoing research work and will be continued in order to develop new garnet materials with improved MO properties for existing and emerging applications in magneto photonic, magneto-plasmonic and integrated optics.

5. Conclusions

Studies of the magneto-optical properties of ultrathin RF magnetron sputtered bismuth-substituted iron garnet films have been conducted in the temperature interval from room temperature down to 8K (−265.15 °C). For the first time, the effects of thin protective Bi_2O_3 layers on the MO properties of ultrathin highly bismuth-substituted dysprosium iron garnet layers have been investigated. At room temperature and also at cryogenic temperatures, the spectral dependencies of magnetic circular dichroism have been measured between 250–850 nm. At room temperature, the MCD spectra typical of Bi-substituted ferrite garnets have been measured (in samples without protective oxide layers), in films of thickness 10.3 nm and above.

In order to improve the MO quality in ultrathin nanocomposite films of co-sputtered composition type $Bi_2Dy_1Fe_4Ga_1O_{12} + Bi_2O_3$, we have introduced a new modification of the annealing crystallization process, in which a 10 nm-thick protective layer of Bi_2O_3 served to prevent the films from Bi content loss otherwise occurring during the high-temperature annealing. Results show that the magnitude of MCD signal measured at 450 nm an oxide-protected annealed film of system $Bi_2Dy_1Fe_4Ga_1O_{12} + Bi_2O_3$, is 2.7 times higher than that of a $Bi_2Dy_1Fe_4Ga_1O_{12}$ unprotected by an oxide layer. The properties of the garnet thin films can also be improved by employing the oxygen plasma treatment just after the deposition process. Additionally, we believe that pre-diffusing the oxygen from plasma into garnet films volume prior to annealing could lead to better (faster) compensation of oxygen loss occurring during sputtering, thus preventing the excessive formation of non-garnet material phases during annealing. However, this preliminary finding may require further experimental confirmation.

Author Contributions: Conceptualization, V.A.K.; M.N.-E.-A. and M.V.; methodology, V.A.K.; M.N.-E.-A. and M.V.; software, M.V.; M.N.-E.-A. and K.A.; validation, V.A.K.; M.N.-E.-A.; M.V.; D.E.B.; V.I.B. and K.A.; formal analysis, V.A.K.; M.N.-E.-A. and M.V.; investigation, V.A.K.; M.N.-E.-A. and M.V.; data curation, V.A.K.; M.N.-E.-A. and M.V.; writing—original draft preparation, V.A.K.; M.N.-E.-A. and M.V.; writing—review and editing, V.A.K.; M.N.-E.-A.; M.V.; D.E.B., V.I.B. and K.A.; visualization, V.A.K.; M.N.-E.-A.; M.V.; D.E.B.; V.I.B. and K.A. All authors have read and agreed to the published version of the manuscript.

Funding: This research is partially supported by the Electron Science Research Institute, Edith Cowan University, Australia. Also this study is supported by a grant from the Russian Foundation for Basic Research (project RFBR 19-07-00444 A) at the Institute of Radio Engineering and Electronics named after V.A. Kotel'nikov, Russian Academy of Sciences (the process of crystallization annealing of thin and ultrathin films of ferrite garnets was tested), as well as RFBR grant (project 19-07-00408 A) the Russian Foundation for Basic Research (project 19-07-00444 A) at Moscow Physical-technical institute (Moscow Institute of Physics and Technology (studies of the optical and magneto-optical properties of experimental samples of films of bismuth-containing ferrite garnets). Studies of the magneto-optical properties of ultrathin films of bismuth-containing ferrite garnets on gadolinium-gallium garnet substrates with thicknesses of 0.7, 1.7 and 3.7 nm were carried out with the support of the state budget.

Acknowledgments: We would like to thank the Electron Science Research Institute, Edith Cowan University, Australia for the support we received during our research.

Conflicts of Interest: The authors declare no conflict of interest.

References

1. Zvezdin, A.K.; Kotov, V.A. *Modern Magnetooptics and Magnetooptical Materials*; Institute of Physics Publishing Bristol and Philadelphia: Boca Raton, FL, USA, 1997. [CrossRef]
2. Kahl, S.; Grishin, S.I.; Kharstev, S.I.; Kawano, K.; Abell, J.S. $Bi_3Fe_5O_{12}$ Thin film visualizer. *IEEE. Trans. Mag.* **2001**, *3*, 2457–2459. [CrossRef]
3. Nur-E-Alam, M.; Vasiliev, M.; Alameh, K. High-performance RF-sputtered Bi-substituted iron garnet thin films with almost in-plane magnetization. *Opt. Mater. Express* **2017**, *7*, 676. [CrossRef]
4. Nur-E-Alam, M.; Vasiliev, M.; Kotov, V.A.; Alameh, K. Highly bismuth-substituted, record-performance magneto-optic garnet materials with low coercivity for applications in integrated optics, photonic crystals, imaging and sensing. *Opt. Mater. Express* **2011**, *1*, 413. [CrossRef]
5. Vasiliev, M.; Alam, M.N.-E.; Kotov, V.A.; Alameh, K.; Belotelov, V.I.; Burkov, V.I.; Zvezdin, A.K. RF magnetron sputtered $(BiDy)_3(FeGa)_5O_{12}:Bi_2O_3$ composite garnet-oxide materials possessing record magneto-optic quality in the visible spectral region. *Opt. Express* **2009**, *17*, 19519. [CrossRef] [PubMed]
6. Syvorotka, I.M.; Ubizskii, S.B.; Kucera, M.; Kuhn, M.; Vertesy, Z. Growth and characterization of Bi, Pr-and Bi, Sc-substituted lutetium iron garnet films with planar magnetization for magneto-optic visualization. *J. Phys. D Appl. Phys.* **2001**, *34*, 117–1187. [CrossRef]
7. Zamani, M.; Ghanaatshoar, M. Adjustable magneto-optical isolators with flat-top responses. *Opt. Express* **2012**, *20*, 24522–24535. [CrossRef] [PubMed]
8. Hibiya, T.; Morishige, Y.; Nakashima, J. Growth and Characterization of Liquid-Phase Epitaxial Bi-Substituted Iron Garnet Films for Magneto-Optic Application. *Jpn. J. Appl. Phys.* **1985**, *24*, 1316–1319. [CrossRef]
9. Huang, M.; Zhang, S. A New Bi-substituted Rare-earth Iron Garnet for a Wideband and Temperature-stabilized Optical Isolator. *J. Mater. Res.* **2000**, *15*, 1665–1668. [CrossRef]
10. Zeazjev, M. Magneto-optic iron-garnet thin films for integrated optical applications. *SPIE Newsroom* **2007**. [CrossRef]
11. Anoikin, E.V.; Sides, P.J. Plasma-activated chemical vapor deposition of bismuth-substituted iron garnets for magneto-optical data storage. *IEEE Trans. Magn.* **1995**, *31*, 3239–3241. [CrossRef]
12. Kang, S.; Yin, S.; Adyam, V.; Li, Q.; Zhu, Y. $Bi_3Fe_4Ga_1O_{12}$ Garnet Properties and Its Application to Ultrafast Switching in the Visible Spectrum. *IEEE Trans. Magn.* **2007**, *43*, 3656–3660. [CrossRef]
13. Aichele, T.; Lorenz, A.; Hergt, R.; Görnert, P. Garnet layers prepared by liquid phase epitaxy for microwave and magneto-optical applications—A review. *Cryst. Res. Technol.* **2003**, *38*, 575–587. [CrossRef]
14. Gomi, M.; Tanida, T.; Abe, M. RF sputtering of highly Bi-substituted garnet films on glass substrates for magneto-optic memory. *J. Appl. Phys.* **1985**, *57*, 3888–3890. [CrossRef]
15. Goossens, V.; Wielant, J.; Van Gils, S.; Finsy, R.; Terryn, H. Optical properties of thin iron oxide films on steel. *Surf. Interface Anal.* **2006**, *38*, 489–493. [CrossRef]
16. Van der Zaag, P.J.; Fontijn, W.F.J.; Gaspard, P.; Wolf, R.M.; Brabers, V.A.M.; van de Veerdonk, R.J.M.; van der Heijden, P.A.A. A study of the magneto-optical Kerr spectra of bulk and ultrathin Fe_3O_4. *J. Appl. Phys.* **1996**, *79*, 5936. [CrossRef]
17. Eschenfelder, A. *Magnetic Bubble Technology*; Springer: New York, NY, USA, 1980.
18. Ivanov, V.E.; Kandaurova, G.S.; Svalov, A.V. Amorphous gadolinium-cobalt films as sensitive media for the topographic mapping nonuniform temperature fields. *Tech. Phys.* **1997**, *42*, 823–827. [CrossRef]

19. Kandaurova, G.S.; Sviderskiy, A.; Klin, V.P.; Chany, V.I. Ferrite-garnet films, domain dynamic structure and anger state. *J. Mag. Mag. Magn.* **1995**, *140*, 2135–2136. [CrossRef]
20. Hanna, S.M.; Friedlaender, F.J.; Gunshor, R.L.; Sato, H. Propagation of surface acoustic-waves in magnetic-bubble garnet-films. *IEEE. Trans. Mag.* **1983**, *19*, 1802–1804. [CrossRef]
21. Kim, Y.H.; Kim, J.S.; Kim, S.I.; Levy, M. Epitaxial Growth and Properties of Bi-Substituted Yttrium-Iron-Garnet Films Grown on (111) Gadolinium-Gallium-Garnet Substrates by Using rf Magnetron Sputtering. *J. Korean Phys. Soc.* **2003**, *43*, 400–405.
22. Jongwannasiri, C.; Watanabe, S. Effects of RF Power and Treatment Time on Wettability of Oxygen Plasma-Treated Diamond-like Carbon Thin Films. *Int. J. Chem. Eng. Appl.* **2014**, *5*, 13–16. [CrossRef]
23. Song, B.J.; Hong, K.; Kim, W.-K.; Kim, K.; Kim, S.; Lee, J.-L. Effect of Oxygen Plasma Treatment on Crystal Growth Mode at Pentacene/Ni Interface in Organic Thin-Film Transistors. *J. Phys. Chem. B* **2010**, *114*, 14854–14859. [CrossRef] [PubMed]
24. Faber, H.; Hirschmann, J.; Klaumünzer, M.; Braunschweig, B.; Peukert, W.; Halik, M. Impact of Oxygen Plasma Treatment on the Device Performance of Zinc Oxide Nanoparticle-Based Thin-Film Transistors. *ACS Appl. Mater. Interfaces* **2012**, *4*, 1693–1696. [CrossRef] [PubMed]
25. Vesel, A.; Mozetic, M. Surface functionalization of organic materials by weakly ionized highly dissociated oxygen plasma. *J. Phys. Conf. Ser.* **2009**, *162*, 012015. [CrossRef]
26. Roh, J.W.; Yang, J.S.; Ok, S.H.; Woo, D.H.; Byun, Y.T.; Jhon, Y.M.; Mizumoto, T.; Lee, W.Y.; Lee, S. Low Temperature O_2 Plasma-Assisted Wafer Bonding of InP and a Garnet Crystal for an Optical Waveguide Isolator. *Solid State Phenom.* **2007**, *124–126*, 475–478. [CrossRef]
27. Mehmood, T.; Kaynak, A.; Dai, X.J.; Kouzani, A.; Magniez, K.; Rubin de Celis, D.; Hurren, C.J.; du Plessis, J. Study of oxygen plasma pre-treatment of polyester fabric for improved polypyrrole adhesion. *Mater. Chem. Phys.* **2014**, *143*, 668–675. [CrossRef]
28. Bhoi, B.; Mahender, C.; Venkataramani, N.; Aiyar, R.P.R.C.; Prasad, S. Effect of Oxygen Pressure on the Magnetic Properties of Yttrium-Iron-Garnet Thin Films Made by Pulsed Laser Deposition. *IEEE Magn. Lett.* **2016**, *7*, 1–4. [CrossRef]
29. Yu, D.; Lu, Y.F.; Xu, N.; Sun, J.; Ying, Z.F.; Wu, J.D. Preparation of α-Al_2O_3 thin films by electron cyclotron resonance plasma-assisted pulsed laser deposition and heat annealing. *J. Vac. Sci. Technol. A Vac. Surf. Film.* **2008**, *26*, 380–384. [CrossRef]
30. Nachimuthu, R.K.; Jeffery, R.D.; Martyniuk, M.; Woodward, R.C.; Metaxas, P.J.; Dell, J.M.; Faraone, L. Investigation of Cerium-Substituted Europium Iron Garnets Deposited by Biased Target Ion Beam Deposition. *IEEE Trans. Magn.* **2014**, *50*, 1–7. [CrossRef]
31. Loho, C.; Djenadic, R.; Bruns, M.; Clemens, O.; Hahn, H. Garnet-Type $Li_7La_3Zr_2O_{12}$ Solid Electrolyte Thin Films Grown by CO_2-Laser Assisted CVD for All-Solid-State Batteries. *J. Electrochem. Soc.* **2017**, *164*, A6131–A6139. [CrossRef]
32. Ma, Q.; Ogino, A.; Matsuda, T.; Nagatsu, M. Defect Control of ZnO Nano-particles Fabricated by Pulsed Nd:YAG Laser Ablation. *Trans. Mater. Res. Soc. Jpn.* **2010**, *35*, 611–615. [CrossRef]
33. Kaynak, A.; Mehmood, T.; Dai, X.; Magniez, K.; Kouzani, A. Study of Radio Frequency Plasma Treatment of PVDF Film Using Ar, O_2 and (Ar + O_2) Gases for Improved Polypyrrole Adhesion. *Materials* **2013**, *6*, 3482–3493. [CrossRef] [PubMed]
34. Leitenmeier, S.; Heinrich, A.; Lindner, J.K.N.; Stritzker, B. Growth of epitaxial bismuth and gallium substituted lutetium iron garnet films by pulsed laser deposition. *J. Appl. Phys.* **2006**, *99*, 08M704. [CrossRef]
35. Pandiyaraj, K.N.; Kumar, A.A.; Ramkumar, M.C.; Sachdev, A.; Gopinath, P.; Cools, P.; De Geyter, N.; Morent, R.; Deshmukh, R.R.; Hegde, P.; et al. Influence of non-thermal $TiCl_4$/Ar + O_2 plasma-assisted TiO_x based coatings on the surface of polypropylene (PP) films for the tailoring of surface properties and cytocompatibility. *Mater. Sci. Eng. C* **2016**, *62*, 908–918. [CrossRef] [PubMed]
36. Christen, H.M.; Ohkubo, I.; Rouleau, C.M.; Jellison, G.E., Jr.; Puretzky, A.A.; Geohegan, D.B.; Lowndes, D.H. A laser-deposition approach to compositional-spread discovery of materials on conventional sample sizes. *Meas. Sci. Technol.* **2005**, *16*, 21–31. [CrossRef]
37. Krumme, J.-P.; David, B.; Doormann, V.; Eckart, R.; Rab, G.; Dossεl, O. Growth, morphology and superconductivity of epitaxial $(RE)_1Ba_2Cl_{13}O_{7-\delta}$ films on $SrTiO_3$ and $NdGaO_3$. *Substrates* **1997**, *1*, 55–68.

38. Kotov, V.A.; Balabanov, D.E.; Grigorovich, S.M.; Kozlov, V.I.; Nevolin, V.K. Magnetic and magnetooptical properties of the transition layer in epitaxial bismuth–gallium iron garnet structures. *Sov. Phys. Tech. Phys.* **1986**, *31*, 544–549.
39. Sun, Y.; Song, Y.-Y.; Chang, H.; Kabatek, M.; Jantz, M.; Schneider, W.; Wu, M.; Schultheiss, H.; Hoffmann, A. Growth and ferromagnetic resonance properties of nanometer-thick yttrium iron garnet films. *Appl. Phys. Lett.* **2012**, *101*, 152405. [CrossRef]
40. Nur-E-Alam, M.; Vasiliev, M.; Alameh, K.; Kotov, V.; Demidov, V.; Balabanov, D. YIG: Bi_2O_3 Nanocomposite Thin Films for Magnetooptic and Microwave Applications. *J. Nanomater.* **2015**, *2015*, 1–6. [CrossRef]
41. Balabanov, D.E.; Kotov, V.A.; Shavrov, V.G.; Vasiliev, M.; Alameh, K. Magneto-optical methods for analysis of nanothick magnetodielectric films. *J. Commun. Technol. Electron.* **2017**, *62*, 78–82. [CrossRef]
42. Vasiliev, M.; Alameh, K.; Nur-E-Alam, M. Analysis, Optimization and characterization of magnetic photonic crystal structures and thin-film material layers. *Technologies* **2019**, *7*, 49. [CrossRef]

Publisher's Note: MDPI stays neutral with regard to jurisdictional claims in published maps and institutional affiliations.

© 2020 by the authors. Licensee MDPI, Basel, Switzerland. This article is an open access article distributed under the terms and conditions of the Creative Commons Attribution (CC BY) license (http://creativecommons.org/licenses/by/4.0/).

Article

Hollow Gold-Silver Nanoshells Coated with Ultrathin SiO$_2$ Shells for Plasmon-Enhanced Photocatalytic Applications

Pannaree Srinoi [1], Maria D. Marquez [1], Tai-Chou Lee [2] and T. Randall Lee [1,*]

[1] Department of Chemistry and the Texas Center for Superconductivity, University of Houston, Houston, TX 77204-5003, USA; pannareeps@gmail.com (P.S.); mdmarqu2@gmail.com (M.D.M.)
[2] Department of Chemical and Materials Engineering, National Central University, Jhongli City 32001, Taiwan; taichoulee@ncu.edu.tw
* Correspondence: trlee@uh.edu

Received: 30 September 2020; Accepted: 31 October 2020; Published: 4 November 2020

Abstract: This article details the preparation of hollow gold-silver nanoshells (GS-NSs) coated with tunably thin silica shells for use in plasmon-enhanced photocatalytic applications. Hollow GS-NSs were synthesized via the galvanic replacement of silver nanoparticles. The localized surface plasmon resonance (LSPR) peaks of the GS-NSs were tuned over the range of visible light to near-infrared (NIR) wavelengths by adjusting the ratio of silver nanoparticles to gold salt solution to obtain three distinct types of GS-NSs with LSPR peaks centered near 500, 700, and 900 nm. Varying concentrations of (3-aminopropyl)trimethoxysilane and sodium silicate solution afforded silica shell coatings of controllable thicknesses on the GS-NS cores. For each type of GS-NS, scanning electron microscopy (SEM) and transmission electron microscopy (TEM) images verified our ability to grow thin silica shells having three different thicknesses of silica shell (~2, ~10, and ~15 nm) on the GS-NS cores. Additionally, energy-dispersive X-ray (EDX) spectra confirmed the successful coating of the GS-NSs with SiO$_2$ shells having controlled thicknesses. Extinction spectra of the as-prepared nanoparticles indicated that the silica shell has a minimal effect on the LSPR peak of the gold-silver nanoshells.

Keywords: nanoparticles; silica shells; metal nanoparticles; gold-silver nanoshells; core-shell nanoparticles

1. Introduction

Harvesting solar energy and storing it in the form of a chemical, such as hydrogen, has garnered significant interest for renewable energy because it is the most abundant and free energy source available on earth [1,2]. Practical and sustainable approaches to solar energy include the utilization of photoelectrochemical reactions such as carbon dioxide reduction and water splitting to afford fuels (e.g., methane, methanol, and hydrogen) [3]. Of particular interest is the water-splitting reaction, for which the standard Gibbs free energy to produce hydrogen from water is greater than 237 kJ/mol and is equivalent to the wavelength of light in the range of 500–1100 nm [4]. Importantly, water is transparent and absorbs little or no light at those wavelengths. Therefore, various composites have been aggressively explored for use in water-splitting reactions due to their ability to absorb across the broad range of wavelengths in the solar spectrum [5].

Abundantly available metal oxide materials such as TiO$_2$, ZnO, Al$_2$O$_3$, and Cu$_2$O have been used as semiconductor photoelectrodes due to their photocatalytic properties [6,7]. However, most of the aforementioned metal oxide materials respond most efficiently to UV light, owing to their large bandgap (higher than 3.2 eV), while the bulk of solar radiation reaching the surface of the earth lies in the visible to near-infrared (NIR) regions [8]. Due to the limited range of absorption, there is a significant need to enhance the photocatalytic properties of metal oxides. Enhancement has been achieved by decorating the semiconductors with dyes, quantum dots, or noble metal/plasmonic nanoparticles [1,9].

Noble metal nanoparticles have been employed in various applications due to their unique chemical and physical properties that are typically different from the properties of the bulk material. One of the most interesting properties of noble metal nanoparticles, such as gold, silver, and copper, is the localized surface plasmon resonance (LSPR) [10,11]. Surface plasmon resonance occurs when the frequency of incident light matches the collective oscillation of surface electrons in the conduction band of the metal. Moreover, the LSPR extinction band can be tuned by varying the size, shape, morphology, and composition of the nanoparticles [12].

Coating a dielectric layer on metallic nanoparticles has the advantage of preventing bulk-metal behavior due to physical contact with neighboring particles or aggregation of unmodified metal nanoparticles. In other words, the dielectric shell plays an important role in enhancing the stability of metal nanoparticles. It is important to employ dielectric materials that are chemically inert, especially for plasmonic enhancement applications, because the optical properties of metal nanoparticles strongly depend on their size and shape. Furthermore, the aggregation of metal nanoparticles has a tremendous effect on the LSPR peak position, which leads to a change in the photo-responsive range of the materials. The SiO_2 shell can maintain the original optical properties of the metal nanoparticles under high light intensity [13].

Additionally, the thickness of the SiO_2 interlayer (i.e., the layer found between the noble metal nanoparticle and a photocatalytic material) has been reported to have an effect on enhancing the photocatalytic activity of photocatalysts [9,14]. The Li group reported the photocatalytic efficiency of an Ag@SiO_2@TiO_2 triplex core-shell photocatalyst as the thickness of the SiO_2 interlayer was varied, and the authors found an increase in photocatalytic efficiency as the SiO_2 thickness was decreased [9]. Unfortunately, the use of silver nanoparticles for these types of applications have the major drawback of their photo-responsive range being limited to only the visible region. As an alternative photoresponsive material, bimetallic core-shell nanoparticles (i.e., nanoshells) exhibit strong plasmon resonances shifted to longer wavelengths, compared to the plasmon resonance of the corresponding solid metal nanospheres, due to plasmonic coupling [15,16]. Specifically, gold-silver nanoshells (GS-NSs), due to their tunable LSPR peak in the range of visible to near-IR light, are good candidates to act as photosensitizers for photocatalysts. Furthermore, the Lee group also observed the benefits of having a silica interlayer between GS-NSs and a zinc indium sulfide (ZIS) outer layer when these composite particles were used as photocatalysts for hydrogen generation via water splitting [14]. Mechanistic studies showed that the silica interlayer blocks the transfer of hot electrons from the GS-NSs to ZIS [14]. Compared to GS-NSs with no SiO_2 coating and those coated with 42 nm of SiO_2, the use of a SiO_2 coating 17 nm thick showed the greatest hydrogen production (2.6-fold increase compared to ZIS alone). The enhanced production observed for the sample with the thinnest SiO_2 shell was attributed to a limited distance for the SPR-generated electromagnetic field to be effective, which translates to higher coupling between the ZIS photocatalyst and the GS-NS core. Therefore, a thinner SiO_2 shell in the aforementioned type of photocatalyst can plausibly lead to plasmon-enhanced hydrogen evolution in water-splitting reactions.

A plethora of reports have been published on the synthesis of SiO_2 shells on metal nanoparticles, with the most common route being modified Stöber methods by addition of tetraethoxysilane in ethanol solution [17]. Additionally, for aqueous media, Mulvany and co-workers have reported a method for the formation of thin silica shells [18]. Following this method, thicker SiO_2 shells can be obtained by using the Stöber method after transferring the thin silica-coated particles into ethanol solution [18]. Furthermore, depending on the concentration of the silica precursors, the Stöber method can yield silica shells with a thickness in the range of 20 to 100 nm. Given the thickness ranges the aforementioned methods yield, there are only a limited number of literature examples that report highly reproducible methods to synthesize homogenous ultra-thin SiO_2 shells on nanoparticle surfaces, especially on rough surfaces such as those on hollow gold-silver alloy nanoshells. As a result, we sought out to develop a reproducible method for the synthesis of homogenous ultra-thin SiO_2 shells on GS-NSs. Herein, we describe the synthesis of hollow gold-silver nanoshells via the technique known as galvanic replacement [19] to give three distinct GS-NS samples having LSPR peaks centered at ~500, ~700, and ~900 nm. Each of the three types of synthesized GS-NSs were then coated with SiO_2 having three different thicknesses (~2, ~10, and ~15 nm) by varying the concentration of (3-aminopropyl)trimethoxysilane (APTMS) and sodium silicate in the solution mixture.

2. Materials and Methods

2.1. Materials

All reagents were purchased from the indicated suppliers and used without further purification. Silver nitrate was purchased from Mallinckrodt (Staines-upon-Thames, United Kingdom). Trisodium citrate, potassium iodide (KI), ascorbic acid, (3-aminopropyl)trimethoxysilane (APTMS), and sodium silicate solution were purchased from Aldrich (St. Louis, MO, USA). Potassium carbonate was purchased from J. T. Baker (Phillipsburg, NJ, USA). Hydrogen tetrachloroaurate(III) hydrate was purchased from Strem (Newburyport, MA, USA). Water was purified to a resistivity of 18 $M\Omega.cm$ (Academic Milli-Q Water system; Millipore Corporation (Burlington, MA, USA)). All of the glassware used during the experiments were cleaned in a base bath, followed by aqua regia solution (3:1 $HCl:HNO_3$), and then dried in the oven and cooled prior to use.

2.2. Preparation of Silver Nanoparticles Cores

Silver nanoparticles (Ag NPs) were prepared by modifying the KI-assisted ascorbic acid/citrate reduction protocol [20] and the method of Lee and Meisel [21]. Specifically, 1 mL of a 5 mM aqueous solution of ascorbic acid (AA) was added into 95 mL of boiling water, followed by boiling for an additional 1 min. An aliquot of $AgNO_3$ (0.0167 g, 0.100 mmol) was dissolved in 2 mL of water, and then 2 mL of 1% trisodium citrate solution and 50 µL of 7 µM KI solution were added. The mixture was placed in an ultrasonic bath for 5 min and was injected into the boiling solution of ascorbic acid. The solution was brought to reflux for 1 h at 120 °C while stirring. The color of the solution quickly changed from colorless to yellow and turned yellowish green after 5 min of reaction, consistent with the presence of silver nanoparticles. The solution was allowed to cool to room temperature (rt) and centrifuged at 8000 rpm for 15 min. The nanoparticles were then redispersed in 12.5 mL water.

2.3. Preparation of Hollow Gold-Silver Nanoshells (GS-NSs)

The hollow GS-NSs were prepared by following an adaptation of a previously reported method [19,22] involving the use of a gold salt (K-Au) solution as the reducing agent. Specifically, 0.025 g of K_2CO_3 was added to 100 mL of water. After 5 min of vigorous stirring, 2 mL of 1% $HAuCl_4 \cdot H_2O$ solution was injected. The color of the mixture changed from yellow to colorless after 30 min of reaction. The flask was covered with aluminum foil to shield it from light and was kept in the refrigerator overnight before being used. To deposit gold on the silver nanoparticles and etch their cores [19,22], various amounts of K-Au solution (10–100 mL) were added to 10 mL of silver nanoparticle solution under vigorous stirring. After 5 h, the solution was centrifuged at 7000 rpm for 15 min. The nanoshells were then redispersed in 10 mL of water at a concentration of 1.0×10^{10} particles/mL.

2.4. Preparation of Silica-Coated Hollow Gold-Silver Nanoshells

Silica-coated hollow GS-NSs were prepared by modification of a previously reported synthetic method [23]. Specifically, 2.5 mL of as-prepared GS-NS suspension was diluted to 40 mL and then mixed with 1 mL of 0.6–1.2 mM APTMS under the vigorous stirring. The mixture was heated to 80 °C, followed by the addition of 10–35 µL of sodium silicate solution. The solution was brought to reflux for 3 h under stirring. The solution was allowed to cool to rt and centrifuged at 6000 rpm for 20 min, and the supernatant decanted. The as-prepared nanoparticles were then redispersed in ethanol. This procedure was repeated three times to remove the residual small silica nanoparticles. Finally, the as-prepared nanoparticles were redispersed in 10 mL of ethanol for characterization and application.

2.5. Characterization of SiO_2-Coated Gold-Silver Nanoshells

The size evaluation and imaging of the as-prepared nanoparticles were performed using a LEO-1525 scanning electron microscope (LEO Electron Microscopy Inc., NY, USA)(SEM) operating at an accelerating voltage of 1.5 kV. All samples were prepared by dropping diluted colloidal nanoparticles onto a silicon wafer and drying in an oven for 15 min. To confirm the size and morphology of nanoparticles at a high resolution, a JEM-2000 FX transmission electron microscope (JEOL, Tokyo, Japan)(TEM) operating at an accelerating voltage of 200 kV was used. All samples were deposited on 300 mesh holey carbon-coated copper grids and dried overnight before analysis. UV-Vis spectra were obtained using a Cary 50 scan UV-visible spectrometer (Agilent Technologies, CA, USA) over a wavelength of 200 to 1000 nm to measure the optical properties. The uncoated nanoparticles and SiO_2-coated GS-NSs were suspended in water and ethanol, respectively, for the measurement. Energy-dispersive X-ray spectroscopy (EDX) data were collected by an EDX attached to the focused ion beam (FIB) instrument and by EDX attached to a JEOL-2200 transmission electron microscope (JEOL, Tokyo, Japan)(TEM) operating at an accelerating voltage of 200 kV.

3. Results and Discussion

3.1. Synthesis of the Hollow Gold-Silver Nanoshells

Monodisperse silver nanoparticles having uniformly spherical shapes were synthesized by modifying a KI-assisted ascorbic acid/citrate reduction protocol. The ascorbic acid was used as the predominant reducing reagent of $AgNO_3$, while KI was used to control the growth of Ag NPs via the adsorption on [111] facets [20]. Citrate has a minor role on the reduction of $AgNO_3$ and is mainly used as a stabilizing agent on the newly formed Ag NPs [20]. Using the Ag NPs as starting materials, hollow gold-silver nanoshells were successfully synthesized via galvanic replacement [22]. As shown in Equation (1), Ag can be oxidized to Ag^+, and the gold salts can be reduced to Au^0 when the K-Au solution is added to the silver nanoparticle cores. This reaction occurs because the standard reduction potential of the Au^{3+}/Au pair is higher than that of the Ag^+/Ag pair, as shown in Equations (2) and (3).

$$3Ag + Au^{3+} \rightarrow 3Ag^+ + Au \tag{1}$$

$$Ag^+ + e^- \rightarrow Ag \quad E^0 = 0.80 \text{ V} \tag{2}$$

$$Au^{3+} + 3e^- \rightarrow Au \quad E^0 = 0.99 \text{ V} \tag{3}$$

Silica-coated hollow gold-silver nanoshells were synthesized using a modification of a previously reported method to coat the GS-NSs with ~2, ~10, and ~15 nm silica shells [23]. As shown in Scheme 1, APTMS was used as a surface-directing agent, and sodium silicate (Na_2SiO_3) solution was used as the silica precursor to coat the GS-NSs. The NH_2 groups of APTMS bind to the surface of the GS-NSs and the $Si(OMe)_3$ groups are available for hydrolysis and condensation with Na_2SiO_3 to deposit a silica layer [23].

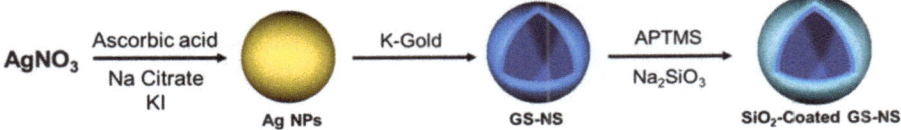

Scheme 1. Synthesis of silica-coated hollow gold-silver nanoshells (GS-NS).

3.2. Morphology and Optical Properties of the Hollow Gold-Silver Nanoshells

The morphologies of the GS-NSs were evaluated using SEM, as shown in Figure 1. Before performing the galvanic replacement with K-Au solution, we determined that the diameter of the silver nanoparticles was in the range of 50–60 nm (Figure 1a). Previous literature has established that addition of K-Au solution to the silver nanoparticles used for their reduction produces small pinholes, and that the pinholes became larger with increasing amounts of K-Au solution [22]. Regardless of the pinhole formation typically observed during the aforementioned process, the GS-NSs retained a diameter of 50–60 nm after the galvanic replacement reaction. The change observed in the morphology of the noble nanoparticles has an effect on their tunable optical properties, which leads to advantages in various kinds of optical applications, including enhanced photocatalysis in a specific region of light, and photothermal treatments and colorimetric sensors [24–26].

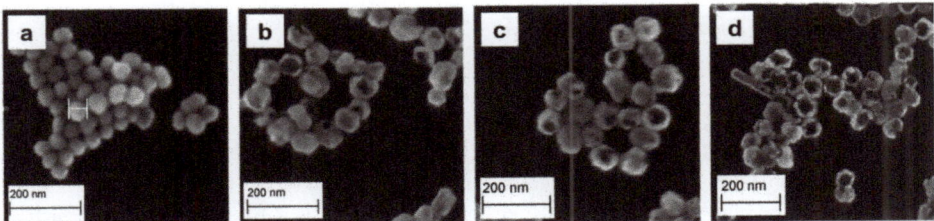

Figure 1. Scanning electron microscopy (SEM) images of (**a**) silver nanoparticles and gold-silver nanoshells synthesized by adding (**b**) 12.5 mL, (**c**) 30 mL, and (**d**) 150 mL of K-Au solution.

3.3. Optical Properties of the Hollow Gold-Silver Nanoshells

Figure 2 illustrates the extinction spectra of silver nanoparticles and GS-NSs prepared using three different amounts of K-Au solution. The uncoated silver nanoparticles exhibit a sharp LSPR peak at ~430 nm. Upon increasing the amount of K-Au solution, the Au shell thickness of the GS-NSs decreases, and the LSPR peak shifts to a longer wavelength. Specifically, the GS-NSs synthesized with a 1:1 ratio of silver nanoparticles to K-Au solution showed an LSPR extinction peak at ~500 nm. Upon increasing the amount of K-Au solution to 1:3 and 1:15 of silver nanoparticles to K-Au solution, the GS-NSs exhibited LSPR extinction peaks at ~700 and ~900 nm, respectively. The LSPR peak is highly tunable in its position and is strongly dependent on the size of the nanoshell as well as shell thickness [22]. Furthermore, shifts in the LSPR position of the GS-NSs also depends on the composition of the nanoparticles. The amount of gold in the GS-NSs is proportional to the amount of K-Au solution added into the silver nanoparticles' suspension. Therefore, adjusting the silver and gold ratios in the GS-NSs can be achieved by varying the amount of K-Au solution, which ultimately can be used to tune the LSPR of the GS-NSs into the visible light to near-IR regions.

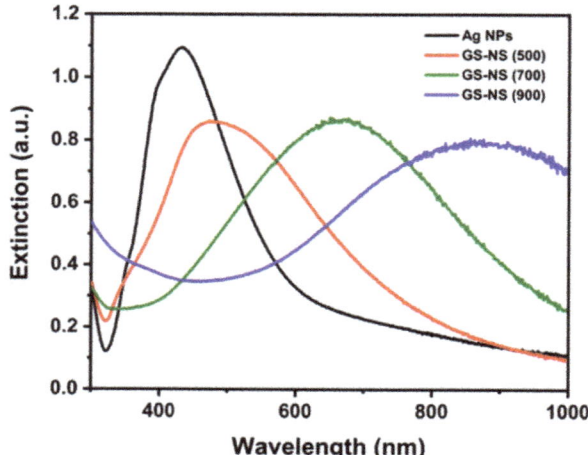

Figure 2. Extinction spectra of Ag NPs and gold-silver nanoshells having LSPR maxima at 500, 700, and 900 nm, denoted as GS-NS (500), GS-NS (700), and GS-NS (900). LSPR = localized surface plasmon resonance

3.4. Composition of the Hollow Gold-Silver Nanoshells

Additionally, TEM-EDX was used to identify the change of silver and gold atomic concentration in the as-prepared GS-NS nanoparticles. As summarized in Table S1 in the Supplementary Materials, the composition of GS-NS (500) showed a 5:1 ratio of Ag/Au. The Ag/Au ratio decreased when the LSPR peak position shifted to a longer wavelength. The Ag/Au ratios of GS-NS (700) and GS-NS (900) decreased to 2:1 and 1:1, respectively. The decrease of the Ag/Au ratio is consistent with the increase of K-Au solution to the silver nanoparticles suspension ratio used in the galvanic replacement reaction. As discussed in the previous section, Au^{3+} in the K-Au solution can be reduced to form Au^0 on the nanoparticles, while the Ag atoms in original silver nanoparticle cores are dissolved in the solution as Ag^+ and etched from the nanoparticles [22].

3.5. Morphology of the SiO$_2$-Coated Gold-Silver Nanoshells

The GS-NSs synthesized with varying amounts of K-Au solution were each further coated with a silica shell. In these experiments, we observed that the silica shell thickness could be controlled by changing the concentration of APTMS and the sodium silicate solution, as shown in Table 1. Controllable silica shells with thicknesses of ~2, ~10, and ~15 nm were successfully coated on GS-NSs using this method. As mentioned above, APTMS acts as a surface-directing agent that binds to the surface of the GS-NSs, while the $Si(OMe)_3$ tails react with the sodium silicate solution via hydrolysis reactions. Therefore, adjusting the concentration of the sodium silicate solution is an important parameter for controlling the thickness of the SiO_2 layer on the surface of the GS-NSs. We note, however, that the concentration of APTMS is also an important factor for controlling the thickness of the SiO_2 shell, as demonstrated by the thin SiO_2 shell obtained from the GS-NS suspension when using low concentrations of APTMS (see Table 1).

Table 1. Concentration and volume of colloidal GS-NS suspension and silica precursors used to obtain silica shells of varying thicknesses.

Nanoparticles	APTMS		Na$_2$SiO$_3$	Silica Shell Thickness (nm)
	Concentration (µM)	Volume (mL)	Volume (µL)	
2.5 mL GS-NS (500)	0.6	1.0	10	2.1 ± 1.0
	1.0	1.0	25	11.5 ± 1.7
	1.5	1.0	35	15.8 ± 1.0
2.5 mL GS-NS (700)	0.6	1.0	10	1.9 ± 1.0
	1.0	1.0	25	9.5 ± 0.9
	1.5	1.0	35	16.0 ± 2.7
2.5 mL GS-NS (900)	0.6	1.0	10	2.2 ± 1.3
	1.0	1.0	25	9.8 ± 0.8
	1.5	1.0	35	15.1 ± 1.0

Figure 3 shows SEM and TEM images of the SiO$_2$-coated hollow GS-NS (900) having three different SiO$_2$ thicknesses (~2, ~10, and ~15 nm). The SEM images indicate that the silica shell was homogenously coated on the surface of the individual GS-NSs. Importantly, the same procedure was applied to the GS-NSs with three different LSPR extinction peaks at 500, 700, and 900 nm to confirm the reproducibility of the synthesis method. As shown in Figure S1 in the Supplementary Materials, the GS-NSs with three different LSPR extinction peak positions (GS-NS (500), GS-NS (700), and GS-NS (900)) were successfully coated with the targeted silica thicknesses using this method. Moreover, as shown in the SEM images, coating with SiO$_2$ shells caused no changes to the original morphologies of the hollow gold-silver nanoshells, which is consistent with the observed negligible effect on the optical properties of the GS-NSs (vide infra). The silica shell thickness distributions of the aforementioned nanoparticles were analyzed by using ImageJ version 1.53d. As summarized in Table 1, the average shell thickness of the SiO$_2$-coated GS-NS (500) nanoparticles synthesized with three different conditions were 2.1 ± 1.0, 11.5 ± 1.7, and 15.8 ± 1.0 nm, respectively. In the case of the SiO$_2$-coated GS-NS (700) nanoparticles, the silica shell thickness was controllable at 1.9 ± 1.0, 9.5 ± 0.9, and 16.0 ± 2.7 nm. Finally, the silica shell thicknesses of the SiO$_2$-coated GS-NS (900) nanoparticles were 2.2 ± 1.3, 9.8 ± 0.8, and 15.1 ± 1.0 nm, respectively.

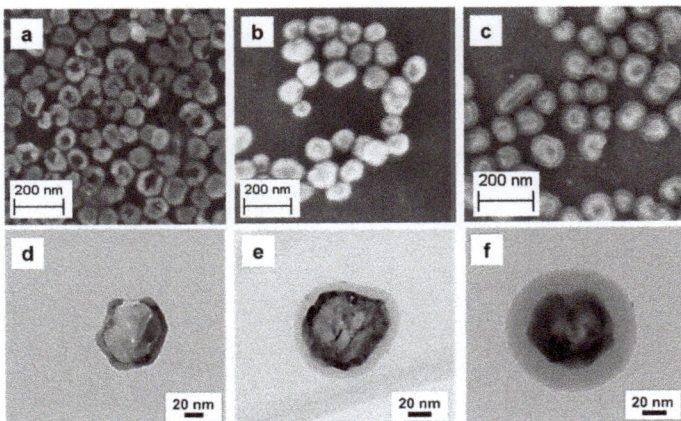

Figure 3. SEM and TEM images of silica-coated hollow gold-silver nanoshells with 900 nm LSPR extinction maxima. The thicknesses of the silica shell are (**a**,**d**) ~2 nm, (**b**,**e**) ~10 nm, and (**c**,**f**) ~15 nm.

We note, however, that due to the limited resolution of the SEM images, the thicknesses of the silica shells are not clearly resolved in the SEM images. Consequently, we used TEM to confirm the thickness of the silica shells on the individual nanoparticles. As a representative example, Figure 3d–f show TEM images of SiO$_2$-coated GS-NS (900) nanoparticles having different silica shell thicknesses. The thicknesses determined from the TEM images were in good agreement with the thicknesses determined from the SEM images (see Table 1 and Figure 3).

To provide further evidence of the silica shell on the GS-NSs following our controllable, reproducible method, TEM-EDX was used. The line spectra, shown in Figure S2 in the Supplementary Materials, confirms the presence of Si arising from the SiO$_2$-shell, validating that silica shells can be grown with controllable thickness simply by changing the concentration of APTMS and sodium silicate in the solution. Supplementary Figure S2b shows the presence of gold, silver, and silicon in the line spectrum of SiO$_2$-coated GS-NSs with a 2 nm SiO$_2$ shell. The blue line of Si corresponding to the SiO$_2$ shell shows ~2 nm Si atop the Ag and Au represented in green and red. The Si layer increases to ~10 to ~15 nm in the line spectra of SiO$_2$-coated GS-NSs having ~10 and ~15 nm SiO$_2$ shell thicknesses, respectively. Additionally, upon the increase of SiO$_2$ thickness, the relative Si magnitude in the TEM line spectra of SiO$_2$-coated GS-NSs having ~10 and ~15 nm shell thicknesses (Supplementary Figure S2d,f) is increased compared to Si in the SiO$_2$-coated GS-NSs having 2 nm shell thicknesses. These results provide further confirmation that the thicknesses of the SiO$_2$ shells can be precisely controlled using our method.

In a recent study, Lee et al. noted that a thin silica interlayer in a GS-NS@SiO$_2$@ZIS nanocomposite enhanced hydrogen production efficiency [14]. The aforementioned nanocomposite was used as a photoelectrode in the photoelectrochemical water-splitting reaction for producing hydrogen gas [14]. As described above, the GS-NS@SiO$_2$@ZIS nanocomposite with a 17 nm silica interlayer exhibited higher hydrogen production efficiency compared to the thicker silica shell (~42 nm) and a sample without a silica interlayer. Specifically, the GS-NS@SiO$_2$@ZIS nanocomposites with a 17 nm SiO$_2$ interlayer showed a hydrogen evolution rate of 0.131 ± 0.03 L/m^2·h, while the GS-NS@SiO$_2$@ZIS nanocomposites with a 42 nm SiO$_2$ interlayer and those with no silica interlayer were approximately 0.07 and 0.09 L/m^2·h, respectively. The SiO$_2$-coated GS-NSs synthesized herein, shown in Figure 3, possess even thinner coatings than the SiO$_2$ interlayers in the nanocomposites evaluated by Lee et al., making them prime candidates for photocatalytic hydrogen production.

3.6. Elemental Composition of the SiO$_2$-Coated Gold-Silver Nanoshells

SEM-EDX data were collected to identify the elemental composition of the SiO$_2$-coated GS-NSs. Figure 4 shows the presence of gold (Mα and Lα peaks at 212 and 971 eV, respectively), silver (Lα peak at 305 eV), and silicon (Kα peak at 174 eV) in the samples. The presence of copper peaks (Lα and Kα peaks at 93 and 894 eV, respectively) arise from the Cu tape used as the substrate in the SEM-EDX samples.

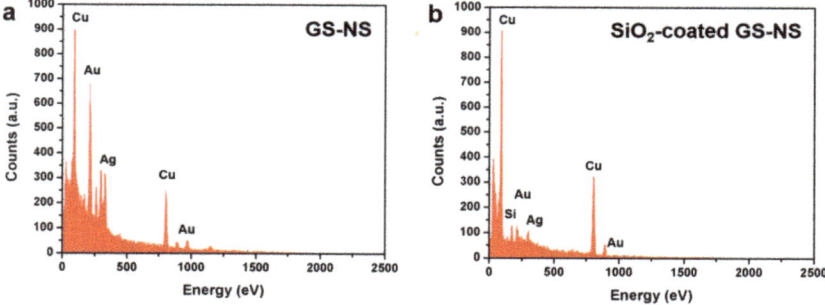

Figure 4. EDX spectra of (**a**) gold-silver nanoshells and (**b**) SiO$_2$-coated gold-silver nanoshells having a 2 nm thick silica shell.

Furthermore, TEM-EDX was used to confirm the atomic concentration of silicon in the as-prepared SiO_2-coated GS-NSs having varied thicknesses. As summarized in Table S2 in the Supplementary Material, the atomic percentage of silicon systematically increased across the series of samples. Specifically, for the SiO_2-coated GS-NS (500) nanoparticles, the atomic percentage of Si increased from 15% for the ~2 nm thick sample to 41% and 71% for the ~10 and ~15 nm thick samples, respectively. The same trend was also observed for the SiO_2-coated GS-NS (700) nanoparticles (i.e., 16%, 57%, and 66%) and the GS-NS (900) nanoparticles (i.e., 13%, 58%, and 75%). These results further confirm that SiO_2-coated GS-NSs with ultrathin silica shells can be successfully synthesized by this method in a controllable manner.

3.7. Optical Properties of the SiO_2-Coated Hollow Gold-Silver Nanoshells

Figure 5 shows the extinction spectra of the as-prepared nanoparticles with LSPR peak maximum at ~900 nm. Upon coating the GS-NSs with silica, the LSPR peak of the nanoparticles slightly shifted to a longer wavelength when compared to the uncoated GS-NSs. Specifically, the GS-NS (900) nanoparticles exhibited an LSPR extinction maximum at 850 nm, which shifted to 868, 870, and 873 nm for the SiO_2-coated GS-NSs having ~2, ~10, and ~15 nm thick SiO_2 shells, respectively. This small shift is likely due to a difference in the refractive index of SiO_2 (1.49) as compared to water (1.33) [23]. We note also that the extinction maxima of the SiO_2-coated GS-NSs having the three different silica thicknesses (~2, 10, and 15 nm) are essentially the same due to the broadness of the LSPR extinction peak (see Figure 5). This result demonstrates that the thickness of the silica shell has a minimal impact on the extinction spectra and thus the optical properties of the gold-silver nanoshells.

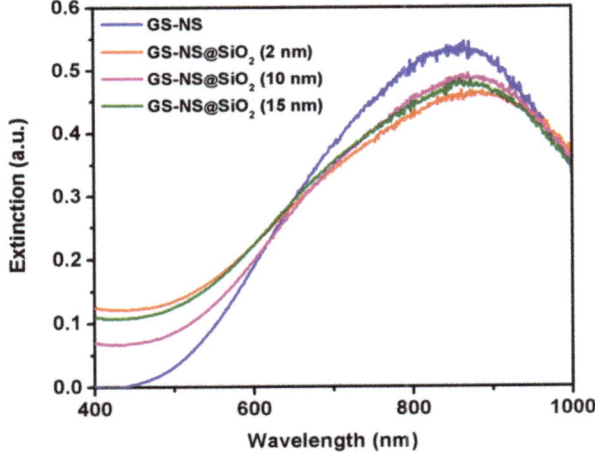

Figure 5. Extinction spectra of SiO_2-coated GS-NS (900) with three different silica shell thicknesses.

4. Conclusions

We have successfully synthesized hollow gold-silver nanoshells coated with silica shells of varying thicknesses by tuning the concentration of (3-aminopropyl)trimethoxysilane and sodium silicate solutions. Our strategy proceeded via the synthesis of silver nanoparticles (cores) in the size range of 60–80 nm using a modified KI-assisted citrate procedure followed by the formation of hollow gold-silver nanoshells via galvanic replacement. Extinction spectra of the GS-NSs demonstrated that these particles are tunable in the visible to NIR region (500–900 nm) by varying the ratio of silver nanoparticles to K-Au solution. Analyses by SEM and TEM were used to establish the size and morphology of the as-prepared nanoparticles and the thickness of the silica coating. The TEM images showed that the silica shell can be obtained at thicknesses of 2, 10, and 15 nm atop the GS-NSs following our method. Importantly, we note that neither of the three obtained SiO_2 shells had detrimental effects on the localized surface plasmon resonance (LSPR) peak of the gold-silver nanoshells. On the whole, the studies described herein demonstrated that SiO_2-coated GS-NSs having ultrathin silica shells are promising materials for plasmon-enhanced photocatalytic applications [9,14].

Supplementary Materials: The following are available online at http://www.mdpi.com/1996-1944/13/21/4967/s1, Table S1: EDX-Derived Composition of GS-NSs with Different LSPR Extinction Peaks, Figure S1: SEM images of silica-coated GS-NSs with different LSPR peak positions. (a–c) GS-NS (500), (d–f) GS-NS (700), and (g–i) GS-NS (900) with ~2, ~10, and ~15 nm, respectively, Figure S2: STEM images and corresponding EDX line scan spectra of SiO_2-coated GS-NSs with (a, b) 2 nm, (c, d) 10 nm, and (e, f) 15 nm silica shell, Table S2: EDX-Derived Composition of the SiO_2-Coated GS-NSs.

Author Contributions: T.R.L. and T.-C.L. were responsible for conceptualization, data curation, funding acquisition, project administration, resources, and supervision; T.R.L., P.S. and M.D.M. were responsible for formal analysis, visualization, writing, and review and editing. P.S. was responsible for the investigation, and T.R.L. and P.S. were responsible for the methodology. All authors have read and agreed to the published version of the manuscript.

Funding: This research was funded by the US Air Force Office of Scientific Research (FA9550-20-1-0349 (20RT0302) and FA9550-18-1-0094), the Robert A. Welch Foundation (E-1320), and the Texas Center for Superconductivity.

Acknowledgments: The authors would like to thank Irene Rusakova and Dezhi Wang for aiding with the TEM measurements.

Conflicts of Interest: The authors declare no conflict of interest.

References

1. Valenti, M.; Jonsson, M.P.; Biskos, G.; Schmidt-Ott, A.; Smith, W.A. Plasmonic Nanoparticle-Semiconductor Composites for Efficient Solar Water Splitting. *J. Mater. Chem. A* **2016**, *4*, 17891–17912.
2. Tuller, H.L. Solar to Fuels Conversion Technologies: A Perspective. *Mater. Renew. Sustain. Energy* **2017**, *6*, 3. [PubMed]
3. Izumi, Y. Recent Advances in the Photocatalytic Conversion of Carbon Dioxide to Fuels with Water and/or Hydrogen Using Solar Energy and Beyond. *Coord. Chem. Rev.* **2013**, *257*, 171–186.
4. Clarizia, L.; Russo, D.; Di Somma, I.; Andreozzi, R.; Marotta, R. Hydrogen Generation through Solar Photocatalytic Processes: A Review of the Configuration and the Properties of Effective Metal-Based Semiconductor Nanomaterials. *Energies* **2017**, *10*, 1624.
5. Fan, W.; Zhang, Q.; Wang, Y. Semiconductor-Based Nanocomposites for Photocatalytic H_2 Production and CO_2 Conversion. *PCCP* **2013**, *15*, 2632–2649. [PubMed]
6. Kawasaki, S.; Takahashi, R.; Yamamoto, T.; Kobayashi, M.; Kumigashira, H.; Yoshinobu, J.; Komori, F.; Kudo, A.; Lippmaa, M. Photoelectrochemical Water Splitting Enhanced by Self-Assembled Metal Nanopillars Embedded in an Oxide Semiconductor Photoelectrode. *Nat. Commun.* **2016**, *7*, 11818.
7. Concina, I.; Ibupoto, Z.H.; Vomiero, A. Semiconducting Metal Oxide Nanostructures for Water Splitting and Photovoltaics. *Adv. Energy Mater.* **2017**, *7*, 1700706.
8. Medina, I.; Newton, E.; Kearney, M.R.; Mulder, R.A.; Porter, W.P.; Stuart-Fox, D. Reflection of near-Infrared Light Confers Thermal Protection in Birds. *Nat. Commun.* **2018**, *9*, 3610.
9. Zhang, X.; Zhu, Y.; Yang, X.; Wang, S.; Shen, J.; Lin, B.; Li, C. Enhanced Visible Light Photocatalytic Activity of Interlayer-Isolated Triplex Ag@SiO_2@TiO_2 Core–Shell Nanoparticles. *Nanoscale* **2013**, *5*, 3359–3366.
10. Li, J.-F.; Zhang, Y.-J.; Ding, S.-Y.; Panneerselvam, R.; Tian, Z.-Q. Core-Shell Nanoparticle-Enhanced Raman Spectroscopy. *Chem. Rev.* **2017**, *117*, 5002–5069.
11. Petryayeva, E.; Krull, U.J. Localized Surface Plasmon Resonance: Nanostructures, Bioassays and Biosensing—A Review. *Anal. Chim. Acta* **2011**, *706*, 8–24. [PubMed]
12. Khan, I.; Saeed, K.; Khan, I. Nanoparticles: Properties, Applications and Toxicities. *Arab. J. Chem.* **2019**, *12*, 908–931.
13. Acharya, D.; Mohanta, B. Optical Properties of Synthesized Ag and Ag@SiO_2 Core-Shell Nanoparticles. *AIP* **2017**, *1832*, 050155.

14. Li, C.-H.; Li, M.-C.; Liu, S.-P.; Jamison, A.C.; Lee, D.; Lee, T.R.; Lee, T.-C. Plasmonically Enhanced Photocatalytic Hydrogen Production from Water: The Critical Role of Tunable Surface Plasmon Resonance from Gold–Silver Nanoshells. *ACS Appl. Mater. Interfaces* **2016**, *8*, 9152–9161.
15. Wang, X.; Feng, J.; Bai, Y.; Zhang, Q.; Yin, Y. Synthesis, Properties, and Applications of Hollow Micro-/Nanostructures. *Chem. Rev.* **2016**, *116*, 10983–11060.
16. Li, C.-H.; Jamison, A.C.; Rittikulsittichai, S.; Lee, T.-C.; Lee, T.R. In Situ Growth of Hollow Gold–Silver Nanoshells within Porous Silica Offers Tunable Plasmonic Extinctions and Enhanced Colloidal Stability. *ACS Appl. Mater. Interfaces* **2014**, *6*, 19943–19950.
17. Stöber, W.; Fink, A.; Bohn, E. Controlled Growth of Monodisperse Silica Spheres in the Micron Size Range. *J. Colloid Interface Sci.* **1968**, *26*, 62–69.
18. Liz-Marzán, L.M.; Giersig, M.; Mulvaney, P. Synthesis of Nanosized Gold–Silica Core–Shell Particles. *Langmuir* **1996**, *12*, 4329–4335.
19. Lu, X.; Chen, J.; Skrabalak, S.E.; Xia, Y. Galvanic Replacement Reaction: A Simple and Powerful Route to Hollow and Porous Metal Nanostructures. *Proc. Inst. Mech. Eng. N* **2007**, *221*, 1–16.
20. Li, H.; Xia, H.; Wang, D.; Tao, X. Simple Synthesis of Monodisperse, Quasi-Spherical, Citrate-Stabilized Silver Nanocrystals in Water. *Langmuir* **2013**, *29*, 5074–5079.
21. Lee, P.C.; Meisel, D. Adsorption and Surface-Enhanced Raman of Dyes on Silver and Gold Sols. *J. Phys. Chem.* **1982**, *86*, 3391–3395. [CrossRef]
22. Vongsavat, V.; Vittur, B.M.; Bryan, W.W.; Kim, J.-H.; Lee, T.R. Ultrasmall Hollow Gold–Silver Nanoshells with Extinctions Strongly Red-Shifted to the Near-Infrared. *ACS Appl. Mater. Interfaces* **2011**, *3*, 3616–3624. [CrossRef] [PubMed]
23. Lee, S.H.; Rusakova, I.; Hoffman, D.M.; Jacobson, A.J.; Lee, T.R. Monodisperse SnO_2-Coated Gold Nanoparticles Are Markedly More Stable Than Analogous SiO_2-Coated Gold Nanoparticles. *ACS Appl. Mater. Interfaces* **2013**, *5*, 2479–2484. [CrossRef] [PubMed]
24. Muller, O.; Dengler, S.; Ritt, G.; Eberle, B. Size and Shape Effects on the Nonlinear Optical Behavior of Silver Nanoparticles for Power Limiters. *Appl. Opt.* **2013**, *52*, 139–149. [CrossRef] [PubMed]
25. Jaque, D.; Martínez Maestro, L.; del Rosal, B.; Haro-Gonzalez, P.; Benayas, A.; Plaza, J.L.; Martín Rodríguez, E.; García Solé, J. Nanoparticles for Photothermal Therapies. *Nanoscale* **2014**, *6*, 9494–9530. [CrossRef]
26. Yu, L.; Li, N. Noble Metal Nanoparticles-Based Colorimetric Biosensor for Visual Quantification: A Mini Review. *Chemosensors* **2019**, *7*, 53. [CrossRef]

Publisher's Note: MDPI stays neutral with regard to jurisdictional claims in published maps and institutional affiliations.

© 2020 by the authors. Licensee MDPI, Basel, Switzerland. This article is an open access article distributed under the terms and conditions of the Creative Commons Attribution (CC BY) license (http://creativecommons.org/licenses/by/4.0/).

Article

Magneto-Fluorescent Hybrid Sensor CaCO$_3$-Fe$_3$O$_4$-AgInS$_2$/ZnS for the Detection of Heavy Metal Ions in Aqueous Media

Danil A. Kurshanov, Pavel D. Khavlyuk, Mihail A. Baranov, Aliaksei Dubavik, Andrei V. Rybin, Anatoly V. Fedorov and Alexander V. Baranov *

Center of Information Optical Technology, ITMO University, 49 Kronverksky Prospekt, 197101 St. Petersburg, Russia; kurshanov.danil@gmail.com (D.A.K.); khavlyuk.stepnogorsk@gmail.com (P.D.K.); mbaranov@mail.ru (M.A.B.); adubavik@itmo.ru (A.D.); rybin@mail.ifmo.ru (A.V.R.); a_v_fedorov@inbox.ru (A.V.F.)
* Correspondence: a_v_baranov@yahoo.com; Tel.: +7-812-457-1781

Received: 31 July 2020; Accepted: 29 September 2020; Published: 30 September 2020

Abstract: Heavy metal ions are not subject to biodegradation and could cause the environmental pollution of natural resources and water. Many of the heavy metals are highly toxic and dangerous to human health, even at a minimum amount. This work considered an optical method for detecting heavy metal ions using colloidal luminescent semiconductor quantum dots (QDs). Over the past decade, QDs have been used in the development of sensitive fluorescence sensors for ions of heavy metal. In this work, we combined the fluorescent properties of AgInS$_2$/ZnS ternary QDs and the magnetism of superparamagnetic Fe$_3$O$_4$ nanoparticles embedded in a matrix of porous calcium carbonate microspheres for the detection of toxic ions of heavy metal: Co^{2+}, Ni^{2+}, and Pb^{2+}. We demonstrate a relationship between the level of quenching of the photoluminescence of sensors under exposure to the heavy metal ions and the concentration of these ions, allowing their detection in aqueous solutions at concentrations of Co^{2+}, Ni^{2+}, and Pb^{2+} as low as ≈0.01 ppm, ≈0.1 ppm, and ≈0.01 ppm, respectively. It also has importance for application of the ability to concentrate and extract the sensor with analytes from the solution using a magnetic field.

Keywords: magnetic–luminescent structure; hybrid system; ternary quantum dots; magnetic nanoparticles; iron oxide; calcium carbonate microspheres; sensor

1. Introduction

The industrial development has resulted in constantly increasing levels of heavy metal contamination in the environment [1,2]. To reduce environmental pollution, it is necessary to determine heavy metal ions in soil [3,4] and water resources [5,6]. The detection and selective quantitative definition of heavy metal ions in nature-conservation resources or biological samples have been an important research area for a long time. The development and creation of sensors based on nanoparticles (NPs) have experienced significant growth over the past decades [7–9].

Traditional methods for heavy metal analysis include atomic absorption spectrometry [10,11], inductively coupled plasma mass spectrometry [12,13], and different electrochemical analyses, e.g., potentiometric techniques [14]. However, these methods have complex sample preparation processes, low stability, and are not compatible with different environments, which significantly limits their applicability.

An alternative method is optical detection. The optical research via fluorescence or colorimetric is the most convenient and hopeful method due to its simplicity and high sensitivity of detection [15,16]. In this area, the application of ion-sensitive luminescent quantum dots (QDs) has been of special interest because of different photophysical properties, customizable structures, and the ability to bind to various ligands, and the possibility integrates into the various hybrid systems with different functional properties while conserving their luminescence properties [17–19].

Many QDs sensor systems are based on binary cadmium compounds, which are hydrophobic toxic compounds [20–22]. Alternative Cd-free nanomaterials are QDs with ternary compositions based on silver or copper, $AgInS_2$, and $CuInS_2$ QDs, which also possess a set of physicochemical properties [23–26] necessary for creating luminescent sensors of toxic compounds. Their photoluminescent (PL) properties are characterized by high quantum PL yield (QYs) ≥ 50% in the visible region from 400 to 1000 nm in the near-infrared (NIR) region after shell growth with, for example, a ZnS semiconductor [24]. Another distinctive characteristic of ternary QDs is broad PL bands with a full width at half-maximum (FWHM) from 100 nm, which slightly overlaps the QD absorption spectrum (Stokes shifts ≈1 eV [27]), in which there is no sharp first exciton peak [28]. Moreover, ternary QDs show long PL lifetimes of some hundred nanoseconds [27,28]. Previous studies demonstrate the possibility of using $AgInS_2$ and $CuInS_2$ QDs in methods for detecting heavy metal ions [29–31].

In addition to detection, it is also important to remove heavy metal ions from the contaminated environment. One of the cleaning methods is sorption using various mesoporous materials, for example, Ca-based materials [32]. Calcium carbonate ($CaCO_3$) crystals are widely used for the manufacturing of carriers containing various embedded nanoparticles or biologically active compounds [33–35]. Nowadays, these $CaCO_3$ crystals (vaterite form) are one of the most popular approaches for matrix formation due to their easy synthesis [36,37], low dispersion, control over crystal size (in the micrometer and submicrometer range [38]), and spherical shape [39], which makes these crystals an attractive candidate for many applications [38,39].

In this work, we focus on the design of $AgInS_2$-based sensors for the most common toxic heavy metal ions, Co^{2+}, Ni^{2+}, and Pb^{2+}, which can accumulate in the human body and cause acute or chronic diseases [40,41]. The sensor is a hybrid complex based on a matrix of porous $CaCO_3$ microspheres doped by Fe_3O_4 magnetic nanoparticles. The surface of the spheres is covered by a shell of several layers of polyelectrolytes. This system combines the properties of several different components, such as the absorption properties of $CaCO_3$ microspheres [42], the photoluminescent properties of $AgInS_2/ZnS$ QDs, and the magnetic properties of Fe_3O_4 nanoparticles.

2. Materials and Methods

2.1. Materials

All reagents purchased from Sigma-Aldrich, Steinheim, Germany, were used without further purification. In all procedures, deionized Hydrolab water was used.

To synthesize $AgInS_2/ZnS$ QDs, we used indium(III) chloride ($InCl_3$), silver nitrate ($AgNO_3$), zinc(II) acetate dihydrate ($Zn(Ac)_2$), sodium sulfide ($Na_2S·9H_2O$), an aqueous solution of ammonia hydrate (NH_4OH), thioglycolic acid (TGA), and isopropyl alcohol.

To synthesize Fe_3O_4 magnetic nanoparticles, we used (tris(acetylacetonato)iron(III)), triethylene glycol (TEG), tetrahydrofuran (THF).

To synthesize $CaCO_3$ microspheres, we used sodium carbonate (Na_2CO_3), calcium chloride ($CaCl_2$), poly(sodium 4-styrenesulfonate) sodium salt (PSS, Mw = 70 kDa) and poly(allylamine hydrochloride) (PAH, Mw = 70 kDa).

The aqueous solutions of heavy metals were prepared by dissolving the salts of metals (cobalt(II) nitrate, nickel(II) sulfate, lead(II) chloride) in water.

2.2. Methods

2.2.1. Synthesis of Fe_3O_4 Nanoparticles

The magnetic Fe_3O_4 nanoparticles were prepared by mixing iron (III) triacetylacetonate and TEG. In the synthesis, 1 mmol of iron precursor and 24 mL of triethylene glycol were added to a three-necked flask under magnetic stirring. The mixture was degassed at up to 90 °C and kept for 60 min under vacuum. After degassing and flushing with argon, the solution was heated to 275 °C and kept for 2 h under a constant Ar flow. After the synthesis, the NCs were washed with THF by centrifugation and then dissolved in water for storage.

2.2.2. Synthesis of $AgInS_2$/ZnS QDs

For the synthesis of $AgInS_2$ quantum dot cores, 1 mL of $AgNO_3$ water solution (0.1 M), 2 mL of TGA water solution (1.0 M), and 0.2 mL of NH_4OH (5.0 M) water solution were added to 92 mL of water under magnetic stirring and ambient conditions. Then, 0.45 mL of NH_4OH solution (5.0 M) and 0.9 mL of $InCl_3$ water solution (1.0 M) containing 0.2 MHNO_3 were added. After that, the solution changed its color from yellow to colorless. After adding 1 mL of 1.0 MNa_2S water solution (1.0 M), the resultant solution was heated at 95 °C for 30 min by a water bath. For ZnS shell growth on the surface of $AgInS_2$, 1 mL of TGA solution (1.0 M) and 1 mL of $Zn(Ac)_2$ solution (1.0 M) containing 0.01 MHNO_3 was added.

After the synthesis, the $AgInS_2$/ZnS quantum dot solution was cooled and concentrated using rotary evaporation. For the size-selection procedure, the aggregation of quantum dots was initiated by adding 0.5 mL of isopropyl alcohol and subsequent centrifugation at 10,000 rpm for 5 min. The precipitate was separated and marked further as fraction #1. This procedure was repeated until the solution was fully discolored. Here, we used quantum dots with a luminescence peak at ≈600 nm.

2.2.3. Preparation of $CaCO_3$-Fe_3O_4-$AgInS_2$/ZnS (CFA) Fluorescent Sensor

Na_2CO_3 (0.33 M; 700 µL), $CaCl_2·2H_2O$ (0.33 M; 700 µL), and Fe_3O_4 nanoparticles (50 µL concentrated aqueous solution) were added to a round-bottom flask. The resulting solution was stirred for 30 s, after which the resulting spheres were centrifuged for 40 s at 3000 rpm. Then, precipitated $CaCO_3$-Fe_3O_4 microspheres were washed twice with H_2O.

The next step was to extend the shell using the Layer-by-Layer (LbL) method. First, 1 mL of PAH solution (6 mg/mL, 0.5 M NaCl, and pH 6.5) was added to the precipitated $CaCO_3$-Fe_3O_4 spheres. The resulting dispersion was shaken for 10 min. Excess polyelectrolyte was removed by washing with water and centrifuging (30 s at 4000 rpm). The procedure was repeated twice. Then, the procedure was repeated using PSS solution (6 mg/mL, 0.5 M NaCl, and pH 6.5). After triple coating with polymer layers (PAH/PSS/PAH), 100 µL of a QDs stock solution was added to the spheres. The dispersion was shaken for 10 min. The excess QDs, by analogy with polyelectrolytes, were centrifuged and washed with water. After the QDs layer, a double layer of PSS and PAH was applied to the spheres.

The resulting fluorescent sensors based on $CaCO_3$ microspheres were dispersed in water.

2.3. Equipments

A spectrophotometer UV-3600 (Shimadzu, Kyoto, Japan) and spectrofluorometer FP-8200 (Jasco, Tokyo, Japan) were used for recording the absorption and PL spectra of the samples, respectively. The SEM and STEM images of the studied samples were taken with a Merlin (Zeiss, Oberkochen, Germany) scanning electron microscope while a FEI Titan electron microscope operating at a voltage of 300 kV was used for getting the TEM image of Fe_3O_4 nanoparticles. The transmitted light images were obtained with a LSM-710 (Zeiss, Oberkochen, Germany) laser scanning confocal microscope equipped with a microobjective of NA = 0.95. The photoluminescence (PL) decay curves of the sample were obtained with a laser scanning confocal microscope MicroTime 100 (PicoQuant, Berlin, Germany) equipped with a pulsed light source with the 80 ps pulses at a repetition rate of 0.2 MHz and a Time Correlated Single Photon Counter (TCSPC) detector. The size of magnetic nanoparticles was determined by dynamic light scattering (DLS) using a Malvern Zetasizer Nano (Malvern, Worcestershire, UK).

3. Results

The formation of the CFA fluorescent sensor was produced (made) in two steps (scheme Figure 1a). In the first step, the $CaCO_3$-Fe_3O_4 microspheres have been formed by using concentrated aqueous solutions of Na_2CO_3 and $CaCl_2$ and magnetic Fe_3O_4 nanoparticles of 5–6 nm mean size. Magnetic Fe_3O_4 nanoparticles are located inside the pores of $CaCO_3$ spheres providing a brownish color of $CaCO_3$-Fe_3O_4. The SEM (STEM) images of $CaCO_3$ microspheres, doped by Fe_3O_4 nanoparticles, Fe_3O_4 magnetic nanoparticles, and $AgInS_2$/ZnS QDs stabilized with TGA are shown in Figure 1b–d, respectively.

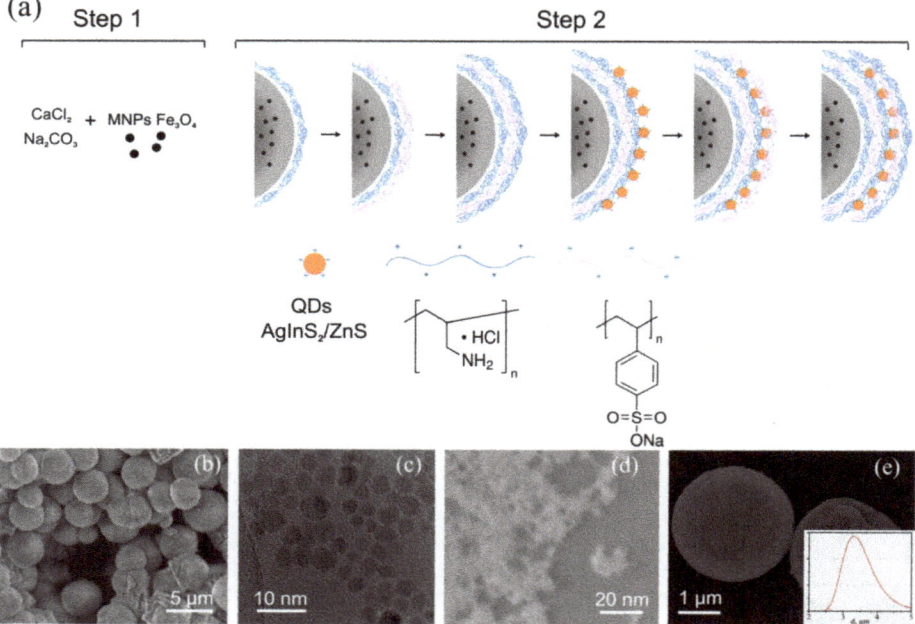

Figure 1. (a) Scheme of a two-stage synthesis of a magnetic luminescent sensor based on $CaCO_3$ microspheres, (b–e) SEM and STEM images of the sensor components: (b) $CaCO_3$ microspheres, doped by Fe_3O_4 nanoparticles, (c) Fe_3O_4 magnetic nanoparticles, (d) $AgInS_2$/ZnS quantum dots (QDs) stabilized with thioglycolic acid (TGA), (e) CFA ($CaCO_3$-Fe_3O_4-$AgInS_2$/ZnS) microspheres with polyelectrolyte shell poly(allylamine hydrochloride)/poly(sodium 4-styrenesulfonate) (PAH/PSS). Inset in (e) shows a typical dynamic light scattering (DLS) size-distribution plot.

The resulting CFA microspheres with polyelectrolytic shell PAH/PSS were characterized by scanning electron microscopy, DLS, UV-Vis, and photoluminescent steady-state and transient microscopy. Their SEM images are presented in Figure 1e and show spherical microparticles with a mean size of ≈4 µm. DLS measurements showed that an average hydrodynamic diameter of the $CaCO_3$-Fe_3O_4 microsphere is about 3.5–4 µm as determined from the analysis of the size-distribution profile (Figure 1e, inset).

The next step was the formation of polyelectrolyte multilayers and the deposition of the QDs layer on a surface of the microspheres by LbL method. Oppositely charged electrostatically interacting polymers and QDs have deposited alternately on the surface of microspheres [43–45]. In this work, we used PAH and PSS polymers and QDs $AgInS_2$/ZnS core/shell capped with the hydrophilic ligand TGA. To heighten the PL properties and stability of the $AgInS_2$ cores, the cores were passivated by a protective ZnS shell. The resulting stable solution QDs $AgInS_2$/ZnS core/shell had a mean size of ≈5 nm and with a PL QY 30%. At first, two double layers of PAH/PSS polyelectrolytes were deposited on the surface of

CaCO$_3$-Fe$_3$O$_4$; then, AgInS$_2$/ZnS QDs were applied over the first layers of the polyelectrolytes shell by electrostatic adsorption using the LbL method. Finally, an additional protective PAH/PSS double layer was placed on the microsphere surface. The resulting CFA fluorescent sensor is illustrated in Figure 1e.

The optical and magnetic properties of the CFA microspheres are illustrated in Figure 2.

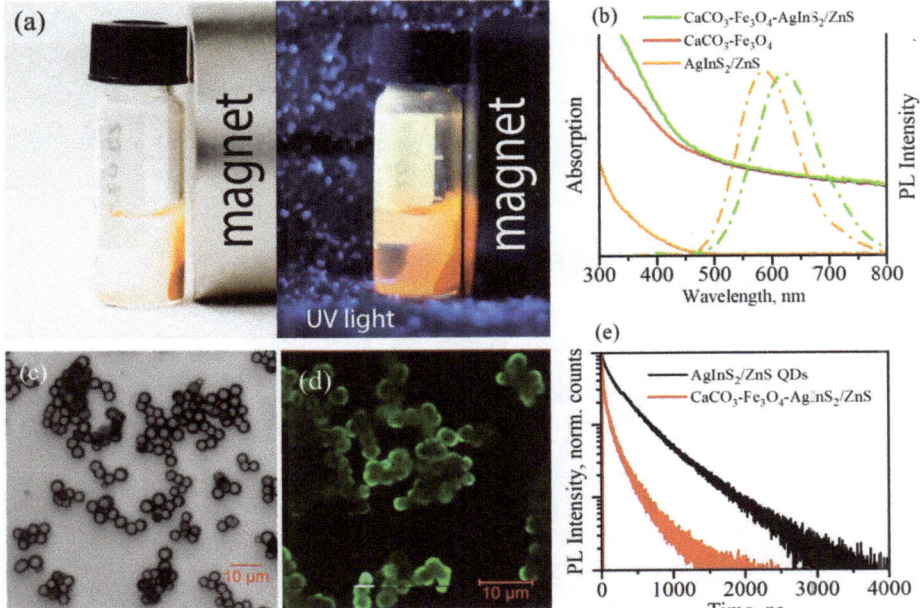

Figure 2. Optical and magnetic properties of the CFA fluorescent sensor. (**a**) Photos of the accumulation of the microspheres to magnetic bar under UV light, (**b**) Absorption (solid line) and photoluminescent (PL) (dash/dotted line) spectra of the CFA fluorescent sensor and AgInS$_2$/ZnS QDs in water (orange AgInS$_2$/ZnS QDs, red CaCO$_3$-Fe$_3$O$_4$ microspheres, and green CFA fluorescent sensor), (**c**) Confocal transmitted light images of the CFA microspheres, (**d**) Fluorescence-Lifetime Imaging Microscopy (FLIM) images of the CFA microspheres, (**e**) PL decay curves of the CFA microspheres and the aqueous dispersion of the initial AgInS$_2$/ZnS QDs. The PL measurements shown in panels (**b**–**e**) were done at 350 nm excitation.

The magneto-fluorescent properties of CFA allow easily concentrating sensors with a magnet or detect their PL under ultraviolet light, as shown in Figure 2a.

The absorption spectra of the AgInS$_2$/ZnS QDs, the CaCO$_3$-Fe$_3$O$_4$ microspheres, and fluorescent CFA sensors are shown in Figure 2b. It is seen that the absorption spectrum of CaCO$_3$-Fe$_3$O$_4$ microspheres contains a noticeable contribution from elastic scattering, which is caused by the rather large size of the microspheres. The absorption of AgInS$_2$/ZnS QDs is characterized by a broad and peak-less spectrum with the absorbance increasing gradually in the shortwave region. The absorption spectra of CaCO$_3$-Fe$_3$O$_4$ microspheres and AgInS$_2$/ZnS QDs additively contribute to the absorption spectrum of the CFA microspheres.

The PL spectra of the CFA microspheres and the AgInS$_2$/ZnS QDs in aqueous solution are shown in Figure 2b. This PL spectrum demonstrates a broad and symmetric luminescence band in the visible and NIR region, similar to that of the AgInS$_2$/ZnS QDs in water. Inclusion of the AgInS$_2$/ZnS QDs into polymer layers on the surface of CaCO$_3$-Fe$_3$O$_4$ microspheres leads to a small redshift of the PL band as compared to that in solution, which is most likely due to the interaction between close QDs. Confocal transmitted light and FLIM images of the CFA microspheres presented in Figure 2c,d, respectively, show that luminescent QDs are embedded in the surface layer of the microspheres of about 3–4 μm in diameter, demonstrating bright PL response.

As in the case of the AgInS$_2$/ZnS QDs in the solution, the PL decay of the QDs in the CFA microspheres is characterized by multiexponential decay kinetics, which is a characteristic property of the PL of ternary QDs that originates from the complicated structure of the low-energy states of the AgInS$_2$ QDs [46]. At the same, an average PL lifetime, calculated by the formula:

$$<\tau> = \Sigma A_i \tau_i^2 / \Sigma A_i \tau_i,$$

where A_i is the amplitude and τ_i is the decay time of the i-th exponent, which for the embedded QDs of about 200 ns is remarkably smaller than that of 350 ns for the QDs in aqueous solution. This difference that is seen in Figure 2e shows that embedding QDs in the surface layer of microspheres results in the appearance of the nonradiative channel of the PL decay.

To demonstrate the ability of our fluorescent sensor for the optical detection of heavy metal ions Co^{2+}, Ni^{2+}, and Pb^{2+}, aqueous metal solutions with a concentration of 0.001 M were prepared. To the CFA solution with a volume of 100 μL per 3 mL of water, metal salt solutions were added with a volume of 3 to 300 μL. Then, the optical properties of the CFA in the presence of heavy metal ions were studied.

The results presented in Figure 3 show that the photoluminescence intensity decreases with an increase in the concentration of heavy metals ions. It is not surprising, since the core–shell AgInS$_2$/ZnS QDs has the stabilizer of thioglycolic acid, which forms the negatively charged layer on the surface of the nanocrystals. The quenching of the PL is observed due to the Coulombic interaction of positively charged metal ions from the analyte with the negative organic shell [47]. Data on the ion concentration dependence of the QD PL are reproduced in three independent experiments, which are reflected in values of corresponding PL intensity measurement errors in the right panels of Figure 3. With low concentrations of Co^{2+} ions, a significant decrease in the luminescence intensity of the sensor is observed (Figure 3a). These suggest a high sensitivity of the fluorescent sensor for Co^{2+} ions. The sensitivity of the sensor for Ni^{2+} and Pb^{2+} was lower by an order of magnitude (Figure 3b,c, respectively). Inserts in Figure 3 (right panel) show in detail the measured PL reduction with the addition of the smallest amount of the Co^{2+}, Ni^{2+}, and Pb^{2+} ions. A simple estimation of low limits of ion concentrations in solution by measuring the reduction of the sensor PL intensity as compared with the ion-free solution within experimental errors of ≈1–2%, gives concentrations of 10^{-8} M Co^{2+} and 10^{-7} M Ni^{2+} and Pb^{2+} even without optimization of the sensor parameters. These values are close to the detection limit reported in [48] for a dissociative CdSe/ZnS QD/PAN complex for luminescent sensing of metal ions in aqueous solutions and orders of magnitude higher for Co(II) and for Ni(II) obtained by colorimetric (absorption) measurements with 1-(2-pyridilazo)-2-naphtol (PAN) as a complexing reagent in the aqueous phase using the non-ionic surfactant Tween 80 [49]. The difference in the luminescence extinguishing of ternary quantum dots by Co^{2+}, Ni^{2+}, and Pb^{2+} is not quite understood at the moment and is the subject of consideration.

At the same time, it should be noted that despite the good sensitivity, the proposed fluorescent sensor does not exhibit selectivity for ions recognition. Therefore, its practical use is limited by a quick on-line preliminary analysis of the presence of heavy metal ions in water samples and a recommendation for further detailed analysis of the elemental composition of ions.

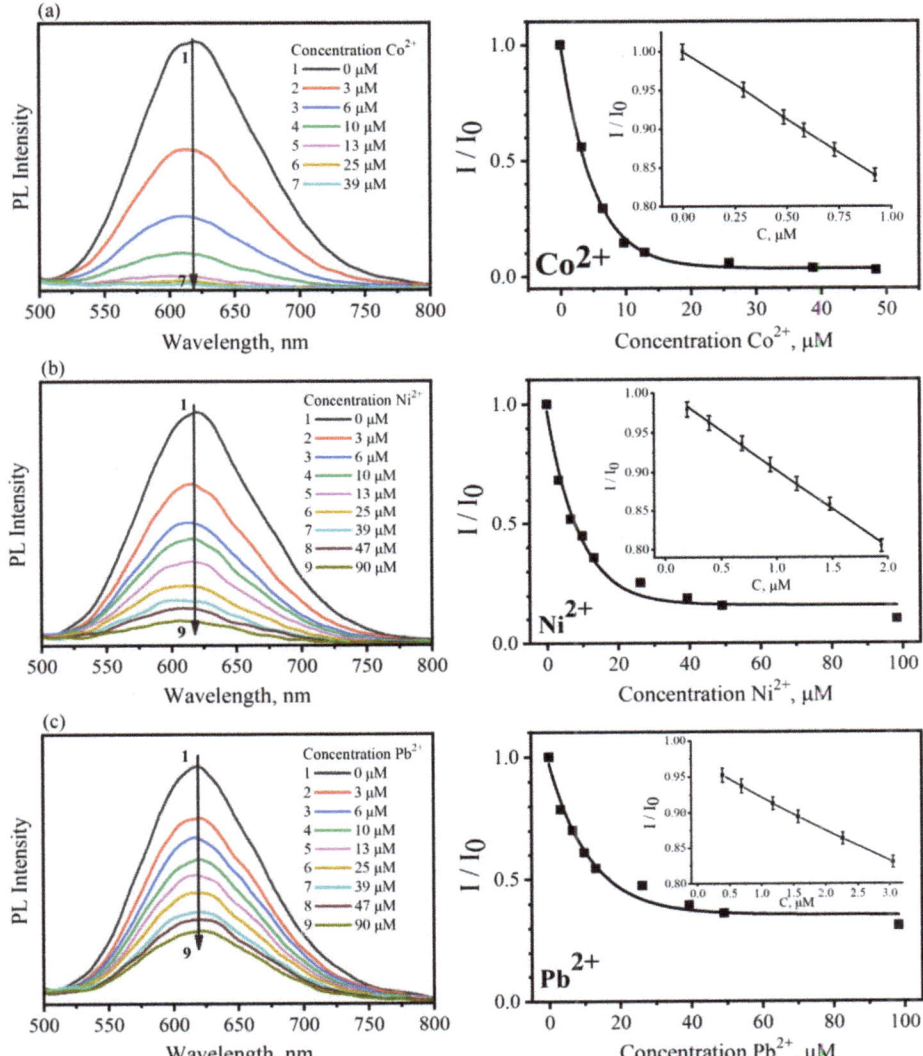

Figure 3. Quenching the CFA sensor luminescence upon increasing concentration heavy metal ions in aqueous solution. Left panel—PL band of CFA sensors as a function of the concentration of the Co^{2+}, Ni^{2+}, and Pb^{2+} ions. Right panel—integrated luminescence intensity ratio (I_0/I) dependence on the concentration of heavy metal ions for (**a**) Co^{2+}, (**b**) Ni^{2+}, (**c**) Pb^{2+}; insets show the regions of low ions concentration.

Due to the presence of magnetic Fe_3O_4 nanoparticles in the microspheres, the sensor with absorbed metal ions can be removed from the solution using a magnetic field, for example, for further analysis of the chemical and/or elemental composition of toxic impurities. To demonstrate this phenomenon, the 20 µL of a 0.001 M aqueous solution of Co^{2+} ions was added to the sensor based on CFA microspheres. In this case, a decrease up to 35% of the initial value was observed in the intensity of the photoluminescence of the sensor (Figure 4b). After the sensor was removed from the test solution using a magnet, the purified solution was again analyzed using the CFA sensor at the same amount (Figure 4c–d). The intensity of luminescence was practically unchanged compared to the solution without metal ions and it decreased by 15%, which indicates a significant decrease in the content of metal ions. This result suggests that the microspheres were successfully absorbed by the heavy metal ions Co^{2+} in solution.

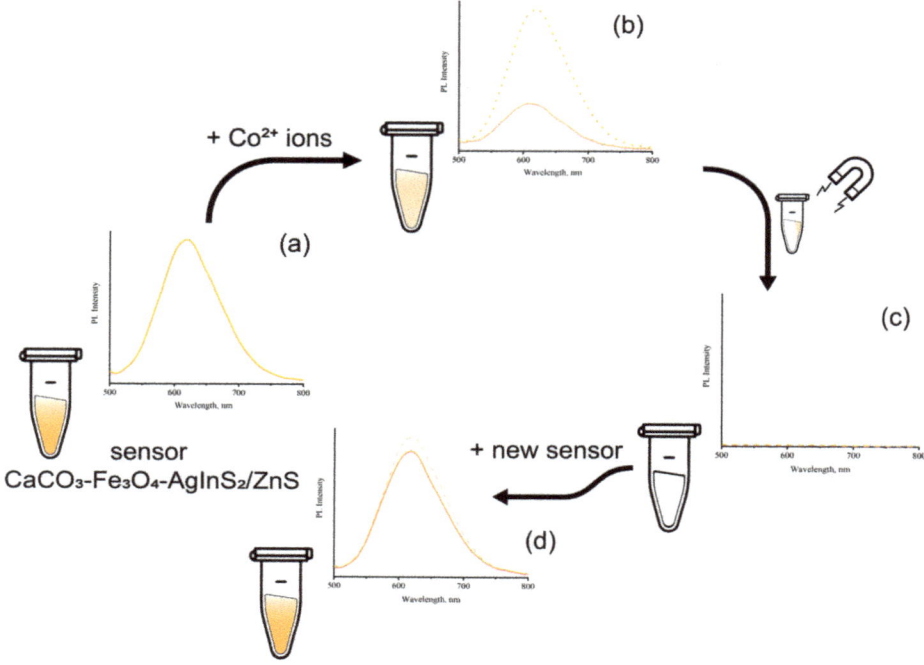

Figure 4. (a) PL spectra of the CFA fluorescent sensor without heavy metal ions, (b) PL spectra of the CFA fluorescent sensor with Co2+ ions in comparison with a solution without the presence of heavy metal ions (dashed line), (c) PL spectra of the solution after removing the CFA fluorescent sensor, (d) PL spectra of the solution after adding a new portion of CFA sensor in comparison with a solution without the presence of heavy metal ions (dashed line).

This result suggests that the microspheres successfully absorb the heavy metal ions in solution, which might be concentrated by the magnetic field e.g., further analytical and (micro)biological applications in flow cytometry.

4. Conclusions

In this work, we obtained a sensor based on a porous matrix, which is microspheres of calcium carbonate doped by magnetic Fe_3O_4 nanoparticles and luminescent $AgInS_2/ZnS$ QDs. The magnetic properties of Fe_3O_4 and photoluminescent properties of QDs $AgInS_2/ZnS$ were combined, yielding highly magnetic and luminescent calcium carbonate microspheres as magneto-fluorescent sensors of heavy metal ions in aqueous solutions. To demonstrate the potential of sensors for sensitivity to in heavy metal ions, the microspheres $CaCO_3$-Fe_3O_4-$AgInS_2/ZnS$ were placed in aqueous solutions containing Co^{2+}, Ni^{2+}, and Pb^{2+} ions with various concentrations. By measuring the quenching the PL of the sensor in the presence of heavy metal ions, we demonstrated the possibility of metal ions detection down to a concentration of 10^{-8} M Co^{2+} and 10^{-7} M Ni^{2+}, and Pb^{2+} ($\approx$0.01 ppm, $\approx$0.1 ppm, and $\approx$0.1 ppm, respectively).

The main distinguishing feature of the proposed sensor from the luminescent sensors previously reported, e.g., [50,51] is the presence of magnetic properties; therefore, in this work, we also demonstrate the ability to concentrate and remove the sensor from the analyzed system together with ions of the analyzed metals for their precise identification, for example, by the inductively coupled plasma mass spectrometry. These concentrations of ions, which could cause a statistically significant quenching of sensors PL, are much lower than their maximum allowable concentration in natural water. It opens prospects of $CaCO_3$-Fe_3O_4-$AgInS_2$/ZnS sensors for developing methods for the environmental monitoring of heavy metal ions.

Author Contributions: Formal analysis, P.D.K.; Investigation, D.A.K. and P.D.K.; Project administration, A.D., A.V.F. and A.V.B.; Visualization, D.A.K. and M.A.B.; Writing—original draft, D.A.K. and A.D.; Writing—review and editing, A.V.B. and A.V.R. All authors have read and agreed to the published version of the manuscript.

Funding: This research was funded by the Federal Target Program for Research and Development of the Ministry of Science and Higher Education of the Russian Federation (Agreement (No. 14.587.21.0047, identifier RFMEFI58718X0047).

Conflicts of Interest: The authors declare no conflict of interest.

References

1. Roy-Chowdhury, A.; Datta, R.; Sarkar, D. Heavy metal pollution and remediation. In *Green Chemistry*; Elsevier: Amsterdam, The Netherlands, 2018; Volume 94, pp. 359–373.
2. Hembrom, S.; Singh, B.; Gupta, S.K.; Nema, A.K. A comprehensive evaluation of heavy metal contamination in foodstuff and associated human health risk: A global perspective. In *Contemporary Environmental Issues and Challenges in Era of Climate Change*; Singh, P., Singh, R.P., Srivastava, V., Eds.; Springer: Singapore, 2020; pp. 33–63. ISBN 978-981-32-9594-0.
3. Dubey, S.; Shri, M.; Gupta, A.; Rani, V.; Chakrabarty, D. Toxicity and detoxification of heavy metals during plant growth and metabolism. *Environ. Chem. Lett.* **2018**, *16*, 1169–1192. [CrossRef]
4. Mallampati, S.R.; Mitoma, Y.; Okuda, T.; Sakita, S.; Kakeda, M. Total immobilization of soil heavy metals with nano-Fe/Ca/CaO dispersion mixtures. *Environ. Chem. Lett.* **2013**, *11*, 119–125. [CrossRef]
5. Taseidifar, M.; Makavipour, F.; Pashley, R.M.; Rahman, A.F.M.M. Removal of heavy metal ions from water using ion flotation. *Environ. Technol. Innov.* **2017**, *8*, 182–190. [CrossRef]
6. Luo, X.; Lei, X.; Cai, N.; Xie, X.; Xue, Y.; Yu, F. Removal of heavy metal ions from water by magnetic cellulose-based beads with embedded chemically modified magnetite nanoparticles and activated carbon. *ACS Sustain. Chem. Eng.* **2016**, *4*, 3960–3969. [CrossRef]
7. Buledi, J.A.; Amin, S.; Haider, S.I.; Bhanger, M.I.; Solangi, A.R. A review on detection of heavy metals from aqueous media using nanomaterial-based sensors. *Environ. Sci. Pollut. Res.* **2020**, 1–9. [CrossRef]
8. Ullah, N.; Mansha, M.; Khan, I.; Qurashi, A. Nanomaterial-based optical chemical sensors for the detection of heavy metals in water: Recent advances and challenges. *TrAC Trends Anal. Chem.* **2018**, *100*, 155–166. [CrossRef]
9. Liu, Y.; Deng, Y.; Dong, H.; Liu, K.; He, N. Progress on sensors based on nanomaterials for rapid detection of heavy metal ions. *Sci. China Chem.* **2017**, *60*, 329–337. [CrossRef]
10. Kunkel, R.; Manahan, S.E. Atomic absorption analysis of strong heavy metal chelating agents in water and waste water. *Anal. Chem.* **1973**, *45*, 1465–1468. [CrossRef] [PubMed]
11. POHL, P. Determination of metal content in honey by atomic absorption and emission spectrometries. *TrAC Trends Anal. Chem.* **2009**, *28*, 117–128. [CrossRef]
12. Bua, D.G.; Annuario, G.; Albergamo, A.; Cicero, N.; Dugo, G. Heavy metals in aromatic spices by inductively coupled plasma-mass spectrometry. *Food Addit. Contam. Part B* **2016**, *9*, 210–216. [CrossRef] [PubMed]
13. Habila, M.A.; ALOthman, Z.A.; El-Toni, A.M.; Soylak, M. Combination of syringe-solid phase extraction with inductively coupled plasma mass spectrometry for efficient heavy metals detection. *CLEAN-Soil Air Water* **2016**, *44*, 720–727. [CrossRef]
14. Privett, B.J.; Shin, J.H.; Schoenfisch, M.H. Electrochemical sensors. *Anal. Chem.* **2010**, *82*, 4723–4741. [CrossRef] [PubMed]

15. Root, H.D.; Thiabaud, G.; Sessler, J.L. Reduced texaphyrin: A ratiometric optical sensor for heavy metals in aqueous solution. *Front. Chem. Sci. Eng.* **2020**, *14*, 19–27. [CrossRef]
16. Rudd, N.D.; Wang, H.; Fuentes-Fernandez, E.M.A.; Teat, S.J.; Chen, F.; Hall, G.; Chabal, Y.J.; Li, J. Highly efficient luminescent metal–organic framework for the simultaneous detection and removal of heavy metals from water. *ACS Appl. Mater. Interfaces* **2016**, *8*, 30294–30303. [CrossRef]
17. Lou, Y.; Zhao, Y.; Chen, J.; Zhu, J.J. Metal ions optical sensing by semiconductor quantum dots. *J. Mater. Chem. C* **2014**, *2*, 595–613. [CrossRef]
18. Devi, P.; Rajput, P.; Thakur, A.; Kim, K.H.; Kumar, P. Recent advances in carbon quantum dot-based sensing of heavy metals in water. *TrAC-Trends Anal. Chem.* **2019**, *114*, 171–195. [CrossRef]
19. Chaniotakis, N.; Buiculescu, R. Semiconductor quantum dots in chemical sensors and biosensors. In *Nanosensors for Chemical and Biological Applications: Sensing with Nanotubes, Nanowires and Nanoparticles*; Woodhead Publishing: Sawston, UK, 2014; ISBN 9780857096609.
20. Bach, L.G.; Nguyen, T.D.; Thuong, N.T.; Van, H.T.T.; Lim, K.T. Glutathione capped cdse quantum dots: synthesis, characterization, morphology, and application as a sensor for toxic metal ions. *J. Nanosci. Nanotechnol.* **2019**, *19*, 1192–1195. [CrossRef]
21. Zhang, K.; Guo, J.; Nie, J.; Du, B.; Xu, D. Ultrasensitive and selective detection of Cu2+ in aqueous solution with fluorescence enhanced CdSe quantum dots. *Sens. Actuators B Chem.* **2014**, *190*, 279–287. [CrossRef]
22. Vázquez-González, M.; Carrillo-Carrion, C. Analytical strategies based on quantum dots for heavy metal ions detection. *J. Biomed. Opt.* **2014**, *19*, 101503. [CrossRef]
23. Raevskaya, A.; Lesnyak, V.; Haubold, D.; Dzhagan, V.; Stroyuk, O.; Gaponik, N.; Zahn, D.R.T.; Eychmüller, A. A fine size selection of brightly luminescent water-soluble Ag–In–S and Ag–In–S/ZnS quantum dots. *J. Phys. Chem. C* **2017**, *121*, 9032–9042. [CrossRef]
24. Jain, S.; Bharti, S.; Bhullar, G.K.; Tripathi, S.K. I-III-VI core/shell QDs: Synthesis, characterizations and applications. *J. Lumin.* **2020**, *219*, 116912. [CrossRef]
25. Stroyuk, O.; Weigert, F.; Raevskaya, A.; Spranger, F.; Würth, C.; Resch-Genger, U.; Gaponik, N.; Zahn, D.R.T. Inherently broadband photoluminescence in Ag–In–S/ZnS quantum dots observed in ensemble and single-particle studies. *J. Phys. Chem. C* **2019**, *123*, 2632–2641. [CrossRef]
26. Cambrea, L.R.; Yelton, C.A.; Meylemans, H.A. ZnS-AgInS2 fluorescent nanoparticles for low level metal detection in water. In *ACS Symposium Series*; American Chemical Society: Washington, USA, 2015; Volume 1210, ISBN 9780841231092.
27. Baimuratov, A.S.; Rukhlenko, I.D.; Noskov, R.E.; Ginzburg, P.; Gun'ko, Y.K.; Baranov, A.V.; Fedorov, A.V. Giant optical activity of quantum dots, rods and disks with screw dislocations. *Sci. Rep.* **2015**, *5*, 14712. [CrossRef] [PubMed]
28. Stroyuk, O.; Raevskaya, A.; Spranger, F.; Selyshchev, O.; Dzhagan, V.; Schulze, S.; Zahn, D.R.T.; Eychmüller, A. Origin and dynamics of highly efficient broadband photoluminescence of aqueous glutathione-capped size-selected Ag–In–S quantum dots. *J. Phys. Chem. C* **2018**, *122*, 13648–13658. [CrossRef]
29. Podgurska, I.; Rachkov, A. Influence of ions of heavy metals on the photoluminescence of nanocrystals AgInS2/ZnS. *Sens. Electron. Microsyst. Technol.* **2017**, *14*, 41–47. [CrossRef]
30. Liu, Y.; Zhu, T.; Deng, M.; Tang, X.; Han, S.; Liu, A.; Bai, Y.; Qu, D.; Huang, X.; Qiu, F. Selective and sensitive detection of copper(II) based on fluorescent zinc-doped AgInS2 quantum dots. *J. Lumin.* **2018**, *201*, 182–188. [CrossRef]
31. Podgurska, I.; Rachkov, A.; Borkovska, L. Effect of Pb 2+ ions on photoluminescence of ZnS-coated AgInS 2 nanocrystals. *Phys. Status Solidi* **2018**, *215*, 1700450. [CrossRef]
32. Mokadem, Z.; Mekki, S.; Saïdi-Besbes, S.; Agusti, G.; Elaissari, A.; Derdour, A. Triazole containing magnetic core-silica shell nanoparticles for Pb 2+, Cu 2+ and Zn 2+ removal. *Arab. J. Chem.* **2017**, *10*, 1039–1051. [CrossRef]
33. Zhao, Y.; Lu, Y.; Hu, Y.; Li, J.-P.; Dong, L.; Lin, L.-N.; Yu, S.-H. Synthesis of superparamagnetic CaCO3 mesocrystals for multistage delivery in cancer therapy. *Small* **2010**, *6*, 2436–2442. [CrossRef]
34. Du, C.; Shi, J.; Shi, J.; Zhang, L.; Cao, S. PUA/PSS multilayer coated CaCO3 microparticles as smart drug delivery vehicles. *Mater. Sci. Eng. C* **2013**, *33*, 3745–3752. [CrossRef]
35. Gusliakova, O.; Atochina-Vasserman, E.N.; Sindeeva, O.; Sindeev, S.; Pinyaev, S.; Pyataev, N.; Revin, V.; Sukhorukov, G.B.; Gorin, D.; Gow, A.J. Use of submicron vaterite particles serves as an effective delivery vehicle to the respiratory portion of the lung. *Front. Pharmacol.* **2018**, *9*. [CrossRef] [PubMed]

36. Sha, F.; Zhu, N.; Bai, Y.; Li, Q.; Guo, B.; Zhao, T.; Zhang, F.; Zhang, J. Controllable synthesis of various CaCO 3 morphologies based on a CCUS idea. *ACS Sustain. Chem. Eng.* **2016**, *4*, 3032–3044. [CrossRef]
37. Parakhonskiy, B.; Zyuzin, M.V.; Yashchenok, A.; Carregal-Romero, S.; Rejman, J.; Möhwald, H.; Parak, W.J.; Skirtach, A.G. The influence of the size and aspect ratio of anisotropic, porous CaCO3 particles on their uptake by cells. *J. Nanobiotechnology* **2015**, *13*, 53. [CrossRef] [PubMed]
38. Wei, W.; Ma, G.-H.; Hu, G.; Yu, D.; Mcleish, T.; Su, Z.-G.; Shen, Z.-Y. Preparation of hierarchical hollow CaCO 3 particles and the application as anticancer drug carrier. *J. Am. Chem. Soc.* **2008**, *130*, 15808–15810. [CrossRef]
39. Boyjoo, Y.; Pareek, V.K.; Liu, J. Synthesis of micro and nano-sized calcium carbonate particles and their applications. *J. Mater. Chem. A* **2014**, *2*, 14270–14288. [CrossRef]
40. Dunnick, J.K.; Elwell, M.R.; Radovsky, A.E.; Benson, J.M.; Hahn, F.F.; Nikula, K.J.; Barr, E.B.; Hobbs, C.H. Comparative carcinogenic effects of nickel subsulfide, nickel oxide, or nickel sulfate hexahydrate chronic exposures in the lung. *Cancer Res.* **1995**, *55*, 5251–5256.
41. Thomson, R.M.; Parry, G.J. Neuropathies associated with excessive exposure to lead. *Muscle Nerve* **2006**, *33*, 732–741. [CrossRef]
42. Fathy, M.; Zayed, M.A.; Moustafa, Y.M. Synthesis and applications of CaCO3/HPC core–shell composite subject to heavy metals adsorption processes. *Heliyon* **2019**, *5*, e02215. [CrossRef]
43. Delcea, M.; Möhwald, H.; Skirtach, A.G. Stimuli-responsive LbL capsules and nanoshells for drug delivery. *Adv. Drug Deliv. Rev.* **2011**, *63*, 730–747. [CrossRef]
44. Marchenko, I.; Yashchenok, A.; Borodina, T.; Bukreeva, T.; Konrad, M.; Möhwald, H.; Skirtach, A. Controlled enzyme-catalyzed degradation of polymeric capsules templated on CaCO3: Influence of the number of LbL layers, conditions of degradation, and disassembly of multicompartments. *J. Control. Release* **2012**, *162*, 599–605. [CrossRef]
45. Martynenko, I.V.; Kusić, D.; Weigert, F.; Stafford, S.; Donnelly, F.C.; Evstigneev, R.; Gromova, Y.; Baranov, A.V.; Rühle, B.; Kunte, H.-J.; et al. Magneto-fluorescent microbeads for bacteria detection constructed from superparamagnetic Fe3O4 nanoparticles and AIS/ZnS quantum dots. *Anal. Chem.* **2019**, *91*, 12661–12669. [CrossRef] [PubMed]
46. Hamanaka, Y.; Ogawa, T.; Tsuzuki, M.; Kuzuya, T. Photoluminescence properties and its origin of AgInS 2 quantum dots with chalcopyrite structure. *J. Phys. Chem. C* **2011**, *115*, 1786–1792. [CrossRef]
47. Chen, J.; Zheng, A.; Gao, Y.; He, C.; Wu, G.; Chen, Y.; Kai, X.; Zhu, C. Functionalized CdS quantum dots-based luminescence probe for detection of heavy and transition metal ions in aqueous solution. *Spectrochim. Acta Part A Mol. Biomol. Spectrosc.* **2008**, *69*, 1044–1052. [CrossRef] [PubMed]
48. Baranov, A.V.; Orlova, A.O.; Maslov, V.G.; Toporova, Y.A.; Ushakova, E.V.; Fedorov, A.V.; Cherevkov, S.A.; Artemyev, M.V.; Perova, T.S.; Berwick, K. Dissociative CdSe/ZnS quantum dot-molecule complex for luminescent sensing of metal ions in aqueous solutions. *J. Appl. Phys.* **2010**, *108*, 074306. [CrossRef]
49. Shar, G.A.; Soomro, G.A. Spectrophotometric determination of cobalt (ii), nickel (ii) and copper (ii) with 1-(2 pyridylazo)-2-naphthol in micellar medium. *Nucleus* **2004**, *41*, 77–82.
50. Mahapatra, N.; Panja, S.; Mandal, A.; Halder, M. A single source-precursor route for the one-pot synthesis of highly luminescent CdS quantum dots as ultra-sensitive and selective photoluminescence sensor for Co 2+ and Ni 2+ ions. *J. Mater. Chem. C* **2014**, *2*, 7373. [CrossRef]
51. Parani, S.; Oluwafemi, O.S. Selective and sensitive fluorescent nanoprobe based on AgInS 2 -ZnS quantum dots for the rapid detection of Cr (III) ions in the midst of interfering ions. *Nanotechnology* **2020**, *31*, 395501. [CrossRef]

© 2020 by the authors. Licensee MDPI, Basel, Switzerland. This article is an open access article distributed under the terms and conditions of the Creative Commons Attribution (CC BY) license (http://creativecommons.org/licenses/by/4.0/).

Article

Hydrothermal Synthesis and Optical Properties of Magneto-Optical Na_3FeF_6:Tb^{3+} Octahedral Particles

Zhiguo Zhao [1,*,†] and Xue Li [2,*,†]

1. Key Laboratory of Electromagnetic Transformation and Detection of Henan province, Luoyang Normal University, Luoyang 471934, China
2. School of Materials Science and Engineering, Zhejiang Sci-Tech University, Xiasha University Town, Hangzhou 310018, China
* Correspondence: zhiguo.zhao@tom.com (Z.Z.); lixue5306@163.com (X.L.)
† Z.Z. and X.L. contributed equally.

Received: 13 December 2019; Accepted: 9 January 2020; Published: 10 January 2020

Abstract: Sodium iron hexafluoride (Na_3FeF_6), as a colorless iron fluoride, is expected to be an ideal host for rare earth ions to realize magneto-optical bi-functionality. Herein, monodispersed terbium ions (Tb^{3+}) doped Na_3FeF_6 particles are successfully synthesized by a facile one-pot hydrothermal process. X-ray diffraction (XRD) and Field emission scanning electron microscopy (FESEM) reveal that the Tb^{3+} doped Na_3FeF_6 micro-particles with regular octahedral shape can be assigned to a monoclinic crystal structure (space group P21/c). Under ultraviolet light excitation, the Na_3FeF_6:Tb^{3+} octahedral particles given orange-red light emission originated from the $^5D_4 \rightarrow {}^7F_J$ transitions of the Tb^{3+} ions. In addition, the magnetism measurement indicates that Na_3FeF_6:Tb^{3+} octahedral particles are paramagnetic with high magnetization at room temperature. Therefore, the Na_3FeF_6:Tb^{3+} powders may find potential applications in the biomedical field as magnetic-optical bi-functional materials.

Keywords: magnetic-optical bi-functional materials; hydrothermal process; down-conversion luminescence; Na_3FeF_6:Tb^{3+}

1. Introduction

Bi-functional materials with distinct magnetic and fluorescent (luminescent) properties have received considerable attention [1,2] due to their potential applications in magnetic resonance imaging [3], targeted drug delivery [4], sensors [5,6], optical isolators [7–9], high accuracy communication [10], and aircraft guidance [11]. To date, there have been a few reports about the synthesis of magnetically-functionalized luminescent materials based on quantum dots (QDs) and organic dyes [12]. However, QDs features notorious disadvantages including chemical instability, potential toxicity, luminescent intermittence and weakly magnetic, while organic dyes typically exhibit rapid photobleaching and a low fluorescence quantum yield [13]. As a result, biological applications of these materials have been seriously restricted.

Rare-earth (RE) ions doped inorganic materials can be considered as alternative luminescent materials in which the above limitations are partly circumvented [14,15]. Nowadays, great efforts have been devoted to the design and fabrication of magneto-optical bi-functional systems based on RE-doped up-conversion or down-conversion materials, such as Gd_2O_3: Er^{3+}/Yb^{3+} [16], $GdPO_4$: Eu^{3+} [17], YVO_4: Er^{3+} [18], $Tb_{0.94}Pr_{0.06}VO_4$ [19] and $NaYF_4$: Yb, Ho [20]. However, studies of magneto-optical effects usually have to rely on materials with a high magnetic movement, which are usually non-transparent. On the other hand, the introduction of strong magnetic (ferromagnetic) materials can be achieved by fabricating magnetic-core/luminescent-shell structures, such as the Fe_3O_4@LaF_3:Yb^{3+}, Er^{3+} [21], Fe_3O_4@α-$NaYF_4$/Yb [13] and Fe_3O_4@ZnO:Er^{3+},Yb^{3+} [22]. However, the preparation processes for core-shell structures is complicated, and more importantly, magnetic oxide, Fe_3O_4, strongly absorbs visible light and quenches fluorescence of the RE ions [23]. Therefore, use of a colorless, strongly magnetic host is of great importance for the development of magneto-optical bifunctional materials.

In this work, colorless Tb^{3+} ions doped sodium iron hexafluoride (Na_3FeF_6:Tb^{3+}) containing a high centration of paramagnetic ion (Fe^{3+}) is synthesized through a simple hydrothermal process. The Na_3FeF_6:Tb^{3+} particles give distinct visible emission under excitation by UV light and its luminescence intensity is optimized by adjusting Tb^{3+} doping concentration. The investigation of the magnetic property reveals that the Na_3FeF_6:Tb^{3+} particles are paramagnetic at room temperature. These results indicate that Na_3FeF_6:Tb^{3+} particles might be promising as a new platform for exploiting magnetic-optical functionalities.

2. Materials and Methods

2.1. Synthesis of Na_3FeF_6:Tb^{3+} Particles

The reagents used in this work were analytical-grade $Fe(NO_3)_3 \cdot 9H_2O$ (99.99%), NH_4HF_2 (99%), NaF (99%), HF (40%), and $Tb(NO_3)_3 \cdot 6H_2O$ (99.99%) (Xiya Reagent, Shandong, China). Samples with a different molar ratio of Tb^{3+} to Fe^{3+} (5%, 10%, 15%, 18%, and 20%) were synthesized by a hydrothermal method under the same conditions. Here, we take Na_3FeF_6:18%Tb^{3+} as an example to present the detailed preparation procedure. The process mainly involves four steps: (1) 14 mL of $Fe(NO_3)_3 \cdot 9H_2O$ solution (0.1 M), 14 mL of NaF solution (0.5 M), 42 mL of NH_4HF_2 solution (0.5 M), and 3 mL of HF were mixed under vigorous magnetic stirring for 30 min; (2) 2.7 mL $Tb(NO_3)_3 \cdot 6H_2O$ was added to the above solution under vigorous magnetic stirring for 3 h; (3) after stirring for 3 h, the above solution was transferred into a Teflon-lined stainless steel autoclave (capacity 100 mL), which was heated at 190 °C for 12 h and cooled naturally to room temperature; (4) the obtained sample was washed by deionized water for several times and dried at 60 °C overnight.

2.2. Characterization

Phase identification of the as-prepared samples were carried out by X-ray diffraction (XRD) (X'Pert Pro, PANalytical BV, Netherland) with Cu Kα radiation (λ = 1.5418 Å). The microstructure and element mapping of particles were observed with a Field emission scanning electron microscopy (FESEM) (Hitachi Ltd., Tokyo, Japan) equipped with an energy dispersive spectroscopy (EDS). UV-Vis (ultraviolet-visible) absorption, transmission, and reflectance spectra of particles were acquired in an UV-Vis spectrophotometer (Model: U3600P) with an integrating sphere using $BaSO_4$ as a standard reference. Photoluminescence excitation and emission spectra were obtained using two spectrometers (Omni-λ3007 and Omni-λ180D; Zolix, Beijing, China) and a 150 W Xenon lamp as the excitation source. The Commission International de l'Eclairage (CIE) chromaticity coordinates of sample were calculated by CIE 1931 software (V.1.6.0.2). Magnetic properties were collected on a Quantum Design superconducting quantum interference device (SQUID) magnetometer (MPMS XL-7).

3. Results and Discussion

Figure 1 shows the typical XRD patterns of the Tb^{3+} doped Na_3FeF_6 samples synthesized with different doping concentrations of Tb^{3+} (5%, 10%, 15%, 18%, and 20%). The diffraction peaks of all samples clearly match that of the standard pattern of Na_3AlF_6 (JCPDS no. 12-0907), indicating the structure of obtained samples is isomorphic with cryolite-like structures (Na_3AlF_6 and Na_3CrF_6) that belongs to the space group P21/c [24–26]. This result agrees with previous report about the structure of Na_3FeF_6 [27]. The three-dimensional crystal structure of Na_3FeF_6:Tb^{3+} is shown in Figure 2. There are three different sodium sites, namely Na1, Na2, and Na3, as highlighted in Figure 2b. Na1 site is located at the distorted octahedral site of (NaF_6), Na2 site is located at the bi-pyramid site of (NaF_5), and the Na3 site is located at the distorted tetrahedral site of (NaF_4). As can be observed from the crystal structure (Figure 2a), Na1 octahedral and Na3 tetrahedral share corners. Na1 octahedral share edges with Na2 bipyramid. Furthermore, all Fe atoms are located at the distorted FeF$_5$ octahedral sites. FeF$_6$ octahedra share corners with Na1 octahedral and Na3 tetrahedral share edges with Na2 bi-pyramid. In this structure, Fe^{3+} sites can be taken by Tb^{3+} ions in Tb doped Na$_3$FeF$_6$. According to the Bragg equation (2dSinθ = nλ), d increases with the decreasing of θ. Figure 1 shows that diffraction peak of Na$_3$FeF$_6$:Tb^{3+} is all shifted to the left compared with that of Na$_3$AlF$_6$ due to the larger ionic size of Fe^{3+} as compared with that of Al^{3+}. As concentration of the Tb ions increases from 5% to 18%, the diffraction peak gradually shifts to the left, diffraction angle θ decreases. This result can be explained by the substitution of Fe^{3+} (ionic radius = 0.65 Å) [28] by Tb^{3+} with a larger ionic radius (0.92 Å) [29]. Therefore, the lattice constant would increase with the increase in the concentration of Tb ions in the lattice. The diffraction peaks of the Na$_3$FeF$_6$ with 18% Tb^{3+} doping are the highest, indicating the best crystallinity. The increase of Tb^{3+} concentration above 18% leads to growth of lattice strain that prevents the further enhancement of crystallization. To further confirm the ions of Tb^{3+} is present in the form of Tb-F and Na$_3$FeF$_6$:18%Tb^{3+} powder was analyzed by XPS (Figure S1). The XPS spectrum shows the presence of Na, F, Fe and Tb elements. Figure S1b shown the XPS spectra of Tb(Ds-4s), Na(2p), F(2s), Fe(3p), Tb(4d) from Na$_3$FeF$_6$:18%Tb^{3+} and the relatively strong peaks at around 7.5, 152 eV can be assigned to the binding energy of Tb (Ds-4s) and Tb (4d), respectively. The peak around 24.7 eV is attributed to the binding energy of Na(2p). The binding energy of F(2s) around 30.3 eV and F(1s) around 684.9 eV are found in spectra of XPS (Figure 1b,d). The Fe(3p) peaks show a doublet around 56.3 and 59.1 eV, corresponding to structure of FeF$_3$ and FeF$_2$, respectively. The result in accordance with the discussion of the XRD patterns of the Tb^{3+} doped Na$_3$FeF$_6$.

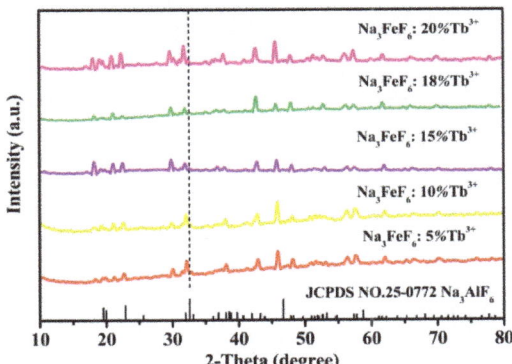

Figure 1. X-ray diffraction (XRD) patterns for samples of the Na$_3$FeF$_6$:Tb^{3+} with different Tb^{3+}-doping concentrations.

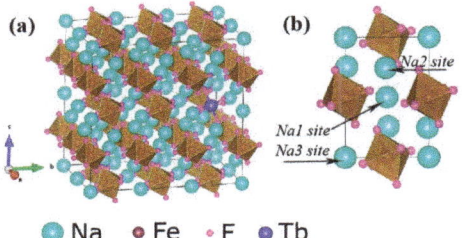

Figure 2. (a) Three-dimensional crystal structure of Na$_3$FeF$_6$:Tb^{3+}. (b) Three different sodium sites in the Na$_3$FeF$_6$ crystal structure.

The Na$_3$FeF$_6$:18%Tb^{3+} particles are then observed by FESEM equipped with an energy dispersive spectroscopy (EDS) device. Figure 3a–c show the FESEM images with low magnification (a) and high magnification (b,c). It can be observed from Figure 3a that the as-prepared samples consist of randomly distributed octahedral particles with a relatively uniform size and shape (edge lengths are approximately 10 µm). As the magnification increases (Figure 3c), it can be seen clearly that the surfaces of the octahedron are almost smooth, but covered by a few small sized particles. EDS analysis was then used to determine the distribution of elements, as illustrated in Figure 3c. The results confirm the dominance of four elements: F, Fe, Na and Tb. In addition, the corresponding EDS mapping images given in Figure 3d–g reveal that all the elements are distribute homogeneously in the particles and Tb ions are successfully doped into the lattice of Na_3FeF_6.

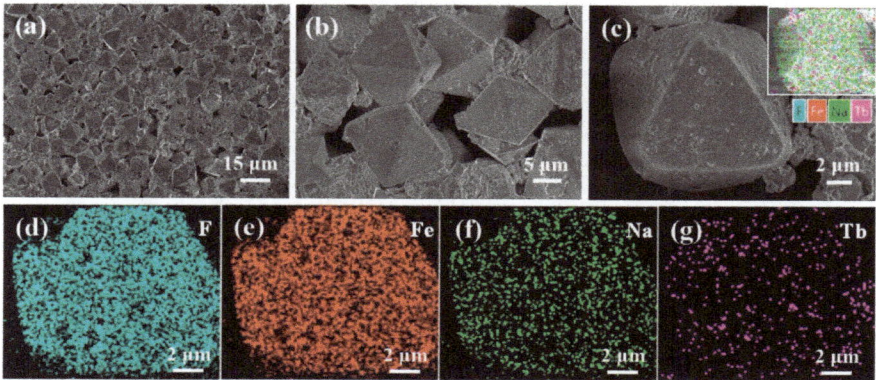

Figure 3. (a–c) Field emission scanning electron microscopy (FESEM) images of the Na_3FeF_6:18%Tb^{3+} powders and (d–g) the corresponding energy dispersive spectroscopy (EDS) mapping for elements image F, Fe, Na, and Tb.

To confirm the optical response of the particles in the UV-Vis range, absorption spectra was detected by an UV-Vis spectrophotometer. As shown in Figure 4, all samples exhibit obvious ultraviolet absorption at wavelength short than 300 nm, which can be attributed to transition of the 4f electronic ground state to the 5d energy levels, namely $^4f_8 \rightarrow ^4f_7{}^5d_1$ energy levels transitions of Tb^{3+} [30]. The f-f transitions of the Tb^{3+} in the wavelength region of 300-400 nm are relatively weak and these peaks at 355 and 380 nm by f-f transitions of Tb^{3+} are almost invisible in the absorption spectra [31]. The transmission spectra and the reflectance spectra of the Na_3FeF_6:18%Tb^{3+} particles correspond to the absorption spectra (as shown in Figure S2).

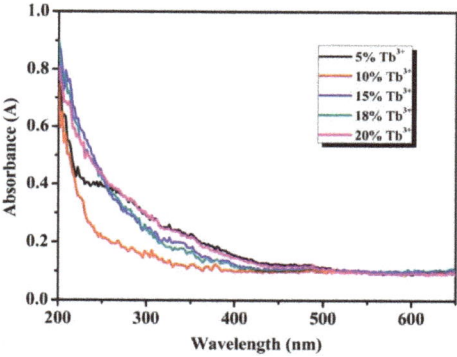

Figure 4. Ultraviolet-visible (UV-Vis) absorption spectra of Na_3FeF_6:Tb^{3+} particles with different Tb^{3+}-doping concentrations.

In order to further study the optical properties of the Na_3FeF_6:Tb^{3+} particles, excitation and emission spectra are measured by fluorescence spectrometers. Figure 5 presents the excitation and emission spectra, and CIE 1931 chromaticity coordinates of the samples together with the energy level diagram of Tb ions. As shown in Figure 5a, the excitation spectra of Na_3FeF_6:18%Tb^{3+} are measured for the emission wavelength of 592 nm. It can be observed that the excitation spectra consist of sharp and intense bands with peak positions at 355 and 375 nm along with weak bands at 280 and 320 nm, which can be assigned to the $^7F_6 \rightarrow {}^5L_{10}$, $^7F_6 \rightarrow {}^5G_6$, $^7F_6 \rightarrow {}^3H_6$ and $^7F_6 \rightarrow {}^5D_1$ transitions of the Tb^{3+}, respectively [32–34].

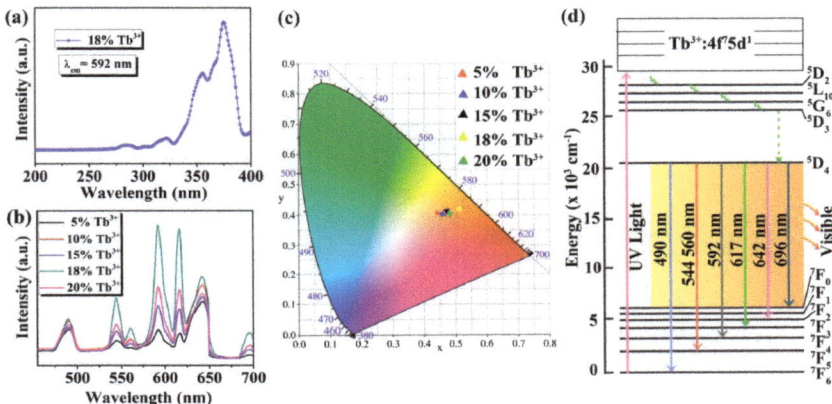

Figure 5. (a) Excitation spectrum of Na_3FeF_6:18%Tb^{3+}; (b) emission spectra of Na_3FeF_6:Tb^{3+} particles with different Tb^{3+}-doping concentrations; (c) CIE 1931 chromaticity diagram of Na_3FeF_6 doped with different concentration of Tb^{3+}; (d) simplified energy levels diagram of the Tb ions.

Since the peak of 375 nm is the strongest in the excitation spectrum, the emission spectra are recorded at this excitation wavelength for Na_3FeF_6:Tb^{3+} particles with different Tb^{3+}-doping concentrations. As can be seen from Figure 5b, the emission spectra in the region of 455–700 nm exhibit seven peaks at 490, 544, 560, 592, 617, 642, and 696 nm due to $^5D_4 \rightarrow {}^7F_6$, $^7F_5$, $^7F_5$, $^7F_4$, $^7F_3$, $^7F_2$ and $^7F_0$ transitions, respectively [35–38]. Among the seven peaks, five peaks at 490, 544, 592, 617, and 642 nm are much stronger, while the other two peaks (560 and 696 nm) are relatively weak. In addition, among the five samples, the luminescence intensity is strongest when the doping concentration of Tb^{3+} is 18%, and the highest luminescence peak is at 592 nm. The emission spectrum is converted to the CIE 1931 chromaticity coordinates using the photoluminescence data to better characterize the emission color of the samples. From the CIE 1931 chromaticity diagram (Figure 5c), it is found that all samples emit orange-red light, which is different from the traditional green light emission of Tb^{3+} ions. This may be due to the use of a new host (Na_3FeF_6) which favors the emission in the longer wavelengths. Furthermore, Figure 5c shows that as the doping concentration of Tb ions increases, the luminescence intensity first increases and then decreases, and the luminescence is strongest at the doping concentration of 18%, which is consistent with the emission spectrum (Figure 5b). The CIE coordinates of Na_3FeF_6:18%Tb^{3+} are X = 0.5103 and Y = 0.4155, which show a typical orange-red color.

In order to better understand the luminescence mechanism of the samples, we combined the energy level diagram of Tb ions (Figure 5d) and take the luminescence at 592 nm as an example to explain the involved electronic transitions. Upon excitation by ultraviolet light (UV-light), Tb^{3+} ions are promoted from the ground state ($^7F_6$) to the excited state (for example $^5L_{10}$, $^5G_6$). Subsequently, the level $^5D_4$ of Tb^{3+} ions is populated by radiation-free transition. Finally, the Tb^{3+} ions relax to the ground state ($^7F_4$) by giving visible emission at around 592 nm. The visible luminescence at other wavelengths is similar to the emission at 592 nm.

Figure 6a shows the temperature-dependence magnetization plots (M-T) in a temperature range between 5 K and 300 K in a 2000 Oe field of Na_3FeF_6:Tb^{3+} particles. It is found that the magnetization decreases rapidly from about 24.74 emu/g at 5 K to 1.06 emu/g at 50 K, and then slowly decreases with a temperature increase from 50 K to 300 K, typical for paramagnetic materials. The magnetization versus magnetic field (M-H) curves at 300 K of Na_3FeF_6:18%Tb^{3+} particles obtained by SQUID magnetometry are presented in Figure 6b. As the strength of the applied magnetic field increasing, the ideal linear correlation between the magnetization and the applied magnetic field was obtained, indicating that Na_3FeF_6:Tb^{3+} possesses paramagnetism. The results show that the synthesized samples might be used as magneto-optical bifunctional materials.

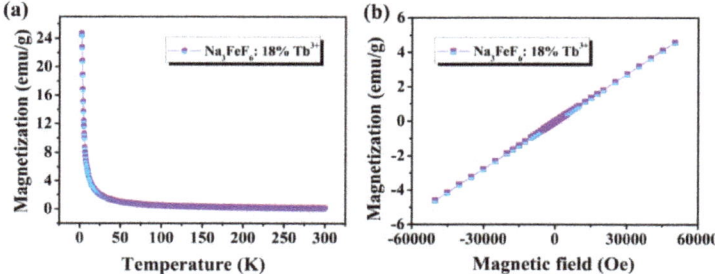

Figure 6. (a) Temperature-dependent magnetization (M-T) curves measured at 2000 Oe for Na_3FeF_6:18%Tb^{3+} particles; (b) magnetization versus magnetic field (M-H) curve at 300 K of Na_3FeF_6:18%Tb^{3+} particles.

4. Conclusions

In summary, monodispersed Na_3FeF_6:Tb^{3+} octahedral particles have been successfully synthesized by a facile one-pot hydrothermal process and the results of XRD and SEM indicated that Na_3FeF_6:Tb^{3+} octahedral belong to a monoclinic crystal structure (space group P21/c). The Na_3FeF_6:Tb^{3+} octahedral particles emit orange-red colored light attributed to the $^5D_4 \rightarrow {}^7F_J$ transitions of the Tb^{3+} ions. The luminescence intensity of the Na_3FeF_6:Tb^{3+} reaches maximum at Tb^{3+} doping concentration of 18%. The M-T and M-H curves confirm that Na_3FeF_6:Tb^{3+} particles are paramagnetic with a high magnetic moment. These results indicate that the obtained Na_3FeF_6:Tb^{3+} octahedral particles might be used as a magnetic-optical bi-functional material for various potential applications in biomedical fields and magneto-optical modulation.

Supplementary Materials: The following are available online at http://www.mdpi.com/1996-1944/13/2/320/s1.

Author Contributions: Z.Z. conceptualization, methodology, data curation and writing-original draft preparation; X.L. preparation of sample, supervision of the work, editing, and review of the manuscript. All authors have read and agreed to the published version of the manuscript.

Funding: This work was financially supported by the National Natural Science Foundation of China (Grant Nos. 61575091, 61675094, and 51471082), excellent Team of Spectrum Technology and Application of Henan province (Grant No. 18024123007).

Conflicts of Interest: The authors declare no conflict of interest.

References

1. Gai, S.; Yang, P.; Li, C.; Wang, W.; Dai, Y.; Niu, N.; Lin, J. Synthesis of Magnetic, Up-Conversion Luminescent, and Mesoporous Core-Shell-Structured Nanocomposites as Drug Carriers. *Adv. Funct. Mater.* **2010**, *20*, 1166–1172. [CrossRef]
2. Xiao, Q.; Zhang, Y.; Zhang, H.; Dong, G.; Han, J.; Qiu, J. Dynamically tuning the up-conversion luminescence of Er^{3+}/Yb^{3+} co-doped sodium niobate nano-crystals through magnetic field. *Sci. Rep.* **2016**, *6*, 31327. [CrossRef] [PubMed]

3. Huang, C.C.; Su, C.H.; Liao, M.Y.; Yeh, C.S. Magneto-optical FeGa$_2$O$_4$ nanoparticles as dual-modality high contrast efficacy T-2 imaging and cathodoluminescent agents. *PCCP* **2009**, *11*, 6331–6334. [CrossRef] [PubMed]
4. Huang, C.C.; Liu, T.Y.; Su, C.H.; Lo, Y.W.; Chen, J.H.; Yeh, C.S. Superparamagnetic hollow and paramagnetic porous Gd$_2$O$_3$ particles. *Chem. Mater.* **2008**, *20*, 3840–3848. [CrossRef]
5. Jung, H.K.; Kim, C.H.; Hong, A.R.; Lee, S.H.; Kim, T.C.; Jang, H.S.; Kim, D.H. Luminescent and magnetic properties of cerium-doped yttrium aluminum garnet and yttrium iron garnet composites. *Ceram. Int.* **2019**, *45*, 9846–9851. [CrossRef]
6. Chen, H.; Qi, B.; Moore, T.; Colvin, D.C.; Crawford, T.; Gore, J.C.; Alexis, F.; Mefford, O.T.; Anker, J.N. Synthesis of brightly PEGylated luminescent magnetic upconversion nanophosphors for deep tissue and dual MRI imaging. *Small* **2014**, *10*, 160–168. [CrossRef]
7. Hu, Q.; Jia, Z.; Yin, Y.; Mu, W.; Zhang, J.; Tao, X. Crystal growth, thermal and optical properties of TSLAG magneto-optical crystals. *J. Alloys Compd.* **2019**, *805*, 496–501. [CrossRef]
8. Li, J.; Tang, T.; Luo, L.; Li, N.; Zhang, P. Spin Hall effect of reflected light in dielectric magneto-optical thin film with a double-negative metamaterial substrate. *Opt. Express* **2017**, *25*, 19117–19128. [CrossRef]
9. Yang, M.; Zhou, D.; Xu, J.; Tian, T.; Jia, R.; Wang, Z. Fabrication and magneto-optical property of yttria stabilized Tb$_2$O$_3$ transparent ceramics. *J. Eur. Ceram. Soc.* **2019**, *39*, 5005–5009. [CrossRef]
10. Ye, S.; Zhang, Y.; He, H.; Qiu, J.; Dong, G. Simultaneous broadband near-infrared emission and magnetic properties of single phase Ni^{2+}-doped β-Ga$_2$O$_3$ nanocrystals via mediated phase-controlled synthesis. *J. Mater. Chem. C* **2015**, *3*, 2886–2896. [CrossRef]
11. Zhang, Y.; Xiao, Q.; He, H.; Zhang, J.; Dong, G.; Han, J.; Qiu, J. Simultaneous luminescence modulation and magnetic field detection via magneto-optical response of Eu^{3+}-doped NaGdF$_4$ nanocrystals. *J. Mater. Chem. C* **2015**, *3*, 10140–10145. [CrossRef]
12. Gu, H.; Zheng, R.; Zhang, X.; Xu, B. Facile one-pot synthesis of bifunctional heterodimers of nanoparticles: A conjugate of quantum dot and magnetic nanoparticles. *J. Am. Chem. Soc.* **2004**, *126*, 5664–5665. [CrossRef] [PubMed]
13. Zhang, F.; Braun, G.B.; Pallaoro, A.; Zhang, Y.; Shi, Y.; Cui, D.; Moskovits, M.; Zhao, D.; Stucky, G.D. Mesoporous multifunctional upconversion luminescent and magnetic "nanorattle" materials for targeted chemotherapy. *Nano Lett.* **2011**, *12*, 61–67. [CrossRef] [PubMed]
14. Jia, H.; Zhou, Y.; Li, X.; Li, Y.; Zhang, W.; Fu, H.; Zhao, J.; Pan, L.; Liu, X.; Qiu, J. Synthesis and phase transformation of NaGdF$_4$: Yb–Er thin films using electro-deposition method at moderate temperatures. *CrystEngComm* **2018**, *20*, 6919–6924. [CrossRef]
15. Jia, H.; Liu, Z.; Liao, L.; Gu, Y.; Ding, C.; Zhao, J.; Zhang, W.; Hu, X.; Feng, X.; Chen, Z. Upconversion Luminescence from Ln^{3+} (Ho^{3+}, Pr^{3+}) Ion-Doped BaCl$_2$ Particles via NIR Light of Sun Excitation. *J. Phys. Chem. C* **2018**, *122*, 9606–9610. [CrossRef]
16. Tan, C.; Ma, B.; Zhang, J.; Zuo, Y.; Zhu, W.; Liu, Y.; Li, W.; Zhang, Y. Pure red upconversion photoluminescence and paramagnetic properties of Gd$_2$O$_3$: Yb^{3+}, Er^{3+} nanotubes prepared via a facile hydrothermal process. *Mater. Lett.* **2012**, *73*, 147–149. [CrossRef]
17. Zhang, L.; Yin, M.; You, H.; Yang, M.; Song, Y.; Huang, Y. Mutifuntional GdPO$_4$: Eu^{3+} hollow spheres: Synthesis and magnetic and luminescent properties. *Inorg. Chem.* **2011**, *50*, 10608–10613. [CrossRef]
18. Ma, Z.W.; Zhang, J.P.; Wang, X.; Yu, Y.; Han, J.B.; Du, G.H.; Li, L. Magnetic field induced great photoluminescence enhancement in an Er^{3+}: YVO$_4$ single crystal used for high magnetic field calibration. *Opt. Lett.* **2013**, *38*, 3754–3757. [CrossRef]
19. Zhu, X.; Tu, H.; Hu, Z.; Zhuang, N. Enhancement of magneto-optical performance of Tb$_{0.94}$Pr$_{0.06}$VO$_4$ single crystals by Pr doping. *Mater. Lett.* **2019**, *242*, 195–198. [CrossRef]
20. Chen, P.; Zhong, Z.; Jia, H.; Zhou, J.; Han, J.; Liu, X.; Qiu, J. Magnetic field enhanced upconversion luminescence and magnetic-optical hysteresis behaviors in NaYF$_4$: Yb, Ho nanoparticles. *RSC Adv.* **2016**, *6*, 7391–7395. [CrossRef]
21. Zhang, L.; Wang, Y.S.; Yang, Y.; Zhang, F.; Dong, W.F.; Zhou, S.Y.; Pei, W.H.; Chen, H.D.; Sun, H.B. Magnetic/upconversion luminescent mesoparticles of Fe$_3$O$_4$@ LaF$_3$: Yb^{3+}, Er^{3+} for dual-modal bioimaging. *Chem. Commun.* **2012**, *48*, 11238–11240. [CrossRef] [PubMed]

22. Peng, H.; Cui, B.; Li, G.; Wang, Y.; Li, N.; Chang, Z.; Wang, Y. A multifunctional β-CD-modified Fe_3O_4@ZnO: Er^{3+}, Yb^{3+} nanocarrier for antitumor drug delivery and microwave-triggered drug release. *Mater. Sci. Eng. C* **2015**, *46*, 253–263. [CrossRef] [PubMed]
23. Zhang, Y.; Pan, S.; Teng, X.; Luo, Y.; Li, G. Bifunctional magnetic- luminescent nanocomposites: Y_2O_3/Tb nanorods on the surface of iron oxide/silica core-shell nanostructures. *J. Phys. Chem. C* **2008**, *112*, 9623–9626. [CrossRef]
24. Brunton, G. The crystal structure of Na_3CrF_6. *Mater. Res. Bull.* **1969**, *4*, 621–626. [CrossRef]
25. Shakoor, R.A.; Lim, S.Y.; Kim, H.; Nam, K.W.; Kang, J.K.; Kang, K.; Choi, J.W. Mechanochemical synthesis and electrochemical behavior of Na_3FeF_6 in sodium and lithium batteries. *Solid State Ion.* **2012**, *218*, 35–40. [CrossRef]
26. Chen, J.Y.; Lin, C.W.; Lin, P.H.; Li, C.W.; Liang, Y.M.; Liu, J.C.; Chen, S.S. Fluoride recovery from spent fluoride etching solution through crystallization of Na_3AlF_6 (synthetic cryolite). *Sep. Purif. Technol.* **2014**, *137*, 53–58. [CrossRef]
27. Jia, H.; Zhou, Y.; Wang, X.; Zhang, W.; Feng, X.; Li, Z.; Fu, H.; Zhao, J.; Liu, Z.; Liu, X. Luminescent properties of Eu-doped magnetic Na_3FeF_6. *RSC Adv.* **2018**, *8*, 38410–38415. [CrossRef]
28. Pang, Y.L.; Abdullah, A.Z. Effect of low Fe^{3+} doping on characteristics, sonocatalytic activity and reusability of TiO_2 nanotubes catalysts for removal of Rhodamine B from water. *J. Hazard. Mater.* **2012**, *235*, 326–335. [CrossRef]
29. Li, X.; Dong, M.; Hu, F.; Qin, Y.; Zhao, L.; Wei, X.; Chen, Y.; Duan, C.; Yin, M. Efficient sensitization of Tb^{3+} emission by Dy^{3+} in $CaMoO_4$ phosphors: Energy transfer, tunable emission and optical thermometry. *Ceram. Int.* **2016**, *42*, 6094–6099. [CrossRef]
30. Bi, F.; Dong, X.; Wang, J.; Liu, G. Electrospinning preparation and photoluminescence properties of $Y_3Al_5O_{12}$: Tb^{3+} nanostructures. *Luminescence* **2015**, *30*, 751–759. [CrossRef]
31. Singh, V.; Singh, N.; Pathak, M.S.; Singh, P.K.; Natarajan, V. Tb^{3+} doped Ca_2La_8 $(SiO_4)_6O_2$ oxyapatite phosphors. *Optik* **2018**, *171*, 356–362. [CrossRef]
32. Kumar, J.S.; Pavani, K.; Sasikala, T.; Jayasimhadri, M.; Jang, K.; Moorthy, L.R. Concentration dependent luminescence characteristics of $^5D_4$ and $^5D_3$ excited states of Tb^{3+} ions in CFB glasses. *Proc. SPIE* **2011**, 79401H. [CrossRef]
33. Cho, I.; Kang, J.G.; Sohn, Y. Photoluminescence imaging of SiO_2@ Y_2O_3: Eu (III) and SiO_2@ Y_2O_3: Tb (III) core-shell nanostructures. *Bull. Korean Chem. Soc.* **2014**, *35*, 575–580. [CrossRef]
34. Wang, D.Y.; Chen, Y.C.; Huang, C.H.; Cheng, B.M.; Chen, T.M. Photoluminescence investigations on a novel green-emitting phosphor Ba_3Sc $(BO_3)_3$: Tb^{3+} using synchrotron vacuum ultraviolet radiation. *J. Mater. Chem.* **2012**, *22*, 9957–9962. [CrossRef]
35. Van Do, P.; Quang, V.X.; Thanh, L.D.; Tuyen, V.P.; Ca, N.X.; Hoa, V.X.; Van Tuyen, H. Energy transfer and white light emission of $KGdF_4$ polycrystalline co-doped with Tb^{3+}/Sm^{3+} ions. *Opt. Mater.* **2019**, *92*, 174–180. [CrossRef]
36. Shi, J.; Wang, Y.; Huang, L.; Lu, P.; Sun, Q.; Wang, Y.; Tang, J.; Belfiore, L.A.; Kipper, M.J. Polyvinylpyrrolidone nanofibers encapsulating an anhydrous preparation of fluorescent SiO_2-Tb^{3+} nanoparticles. *Nanomaterials* **2019**, *9*, 510. [CrossRef] [PubMed]
37. Blasse, G.; Bril, A. Investigations of Tb^{3+}-activated phosphors. *Philips Res. Rep.* **1967**, *22*, 481–504.
38. Scholl, M.S.; Trimmier, J.R. Luminescence of YAG: Tm, Tb. *J. Electrochem. Soc.* **1986**, *133*, 643–648. [CrossRef]

© 2020 by the authors. Licensee MDPI, Basel, Switzerland. This article is an open access article distributed under the terms and conditions of the Creative Commons Attribution (CC BY) license (http://creativecommons.org/licenses/by/4.0/).

Article

Nonlinear Optical Phenomena in a Silicon-Smectic A Liquid Crystal (SALC) Waveguide

Boris I. Lembrikov *, David Ianetz and Yosef Ben-Ezra

Faculty of Electrical Engineering, Holon Institute of Technology, P.O. Box 305, 52 Golomb str., Holon 58102, Israel
* Correspondence: borisle@hit.ac.il; Tel.: +972-3-502-6684

Received: 13 May 2019; Accepted: 22 June 2019; Published: 28 June 2019

Abstract: Liquid crystals (LCs) are organic materials characterized by the intermediate properties between those of an isotropic liquid and a crystal with a long range order. The LCs possess strong anisotropy of their optical and electro-optical properties. In particular, LCs possess strong optical nonlinearity. LCs are compatible with silicon-based technologies. Due to these unique properties, LCs are promising candidates for the development of novel integrated devices for telecommunications and sensing. Nematic liquid crystals (NLCs) are mostly used and studied. Smectic A liquid crystals (SALCs) have a higher degree of long range order forming a layered structure. As a result, they have lower scattering losses, specific mechanisms of optical nonlinearity related to the smectic layer displacement without the mass density change, and they can be used in nonlinear optical applications. We theoretically studied the nonlinear optical phenomena in a silicon-SALC waveguide. We have shown theoretically that the stimulated light scattering (SLS) and cross-phase modulation (XPM) caused by SALC nonlinearity can occur in the silicon-SALC waveguide. We evaluated the smectic layer displacement, the SALC hydrodynamic velocity, and the slowly varying amplitudes (SVAs) of the interfering optical waves.

Keywords: silicon photonics; optical waveguide; smectic A liquid crystal (SALC); stimulated light scattering (SLS)

1. Introduction

Liquid crystals (LCs) are promising candidates for applications in novel integrated devices for telecommunications, sensing, and lab-on-chip bioscience [1]. These applications are based on the unique optical properties of LC. The orientational energy of LC molecules is comparatively small, and for this reason they are characterized by an easy susceptibility to external field perturbation [2]. As a result, the LC effective refractive index can be controlled by an external electric field which may be used for optical transmission, reflection, switching, and modulation applications [2]. LCs are highly nonlinear optical materials because their properties such as temperature, molecular orientation, density, and electronic structure can be easily perturbed by an applied optical field [2].

The liquid crystal on silicon technology (LCOS) is widely used in telecommunications [3,4]. The basic element of the LCOS technology is the LCOS cell consisting of the LCOS backplane, LC layer and cover glass [3,4]. The LCOS cell can simultaneously perform the electrical and optical functions [3,4]. The photonic applications of the LCOS devices include the spatial light modulation (SLM), the holographic beam steering, optical wavelength selective switching, and the optical power control [4]. The LCOS SLM technology is a promising candidate for the so-called structured light where the optical field amplitude, phase, and polarization can be controlled spatially while the time and frequency spectrum can be controlled temporally [3]. Nonlinear silicon photonics can be used in on-chip optical signal processing and computation due to its low cost and compatibility with CMOS technology [5]. The development of nonlinear silicon photonics is limited by the absence of the second-order nonlinear susceptibility $\chi^{(2)}$ due to centrosymmetric structure of Si, comparatively low third-order nonlinear susceptibility $\chi^{(3)}$, the two-photon absorption (TPA) and free carrier absorption (FCA) [5]. To mitigate these disadvantages new materials with better nonlinear properties may be integrated with silicon. In such a case, the new materials may improve the nonlinearity of an optical device while silicon can confine the optical modes to nanoscale [5]. The organic nonlinear materials with a large $\chi^{(3)}$ can be used for the creation of a silicon-organic hybrid waveguide [5]. In particular, liquid crystals (LCs) may be used as a waveguide core where the modulation and switching of photonic signals is possible by using electro-optic or nonlinear optic effects [1,6–13].

We briefly discuss the basic properties of LCs. LCs are characterized by the properties intermediate between solid crystalline and liquid phases [2,14]. LCs flow like liquids, but possess a partial long range order and anisotropy of their physical parameters such as dielectric constants, elastic constants, viscosities, nonlinear susceptibilities [2]. Various phases in which such materials can exist are called mesophases [2]. There are three types of LCs: thermotropic LCs, polymeric LCs, and lyotropic LCs [2,14].

(1) Lyotropic LCs can be obtained in a solution with an appropriate concentration of a material.
(2) Polymeric LCs are the polymers consisting of the monomer LC molecules.
(3) Thermotropic LCs self-assemble in various ordered arrangement of their crystalline axis depending on the temperature.

Thermotropic LCs are most widely used and studied because of their extraordinary linear, electro-optical, and nonlinear optical properties and the possibility to control the transitions between different mesophases by varying the operating temperature [1,2]. The thermotropic LCs consist of elongated molecules with the direction of their axes determined by the unit vector $\vec{n}$ called director [2,14]. The long range ordering of LC mesophase is characterized by the director spatial distribution [2,14]. There exist three main types of thermotropic LCs: nematic LC (NLC), cholesteric LC (CLC), and smectic LC (SLC) [2,14]. NLC molecules are centrosymmetric in such a way that $\vec{n}$ and $-\vec{n}$ are equivalent, the molecules are positionally random, but they are mostly aligned in the direction defined by the director $\vec{n}$ [2,14]. CLC consists of chiral molecules, or they may be obtained by adding of chiral molecules to NLC [2]. As a result, they exhibit a helical structure where the direction $\vec{n}$ of the molecular orientation rotates in space around the helical axis with a period of about 300 nm [2]. The phase transition between the nematic and smectic A phases had been investigated both theoretically and experimentally in a large number of publications (see, for example, [2,15–22]). It is essentially the second kind phase transition [18]. The phase transition temperature T_{SmA-N} may be different for different LC materials. For example, for 8CB $T_{SmA-N} \approx 307$ K, for 9CB $T_{SmA-N} \approx 321$ K [22].

SLC are characterized by the positional long range order in the direction of the elongated molecular axis and possess a layered structure with a layer thickness of about 2 nm approximately equal to the length of a SLC molecule [2,14]. Inside the layers, molecules are not ordered and represent a two-dimensional liquid [2,14]. There are different SLC phases [2,14].

(1) Smectic A LC (SALC) where the long axes of the molecules are perpendicular to the layer plane.
(2) Smectic B LC where the hexagonal in-layer ordering of the molecules perpendicular to the layer plane exists.
(3) Smectic C LC where the molecules are tilted with respect to the layers.
(4) Smectic C* LC consisting of the chiral molecules and possessing the spontaneous polarization.
(5) So-called exotic smectic phases.

The nonlinear optical phenomena such as degenerate and nondegenerate wave mixing, optical bistabilities and instabilities, self-focusing and self-guiding, phase conjugation, stimulated light scattering (SLS), optical limiting, interface switching, beam combining, and self-starting laser oscillations have been observed in liquid crystalline materials [14,23]. NLCs are mainly studied and used in linear and nonlinear optical applications [2,4,23]. For instance, in NLC the optically induced director axis reorientation results in the so-called giant optical nonlinearity (GON) with the nonlinear refractive index coefficient $n_2^{NL} \sim 10^{-4} - 10^{-3}$ cm^2/W [14]. However, NLCs are characterized by large losses and relatively slow responses limiting their integrated electro-optical applications [2]. The light scattering properties of SALC thin film waveguide have been studied both theoretically and experimentally [24]. The scattering losses in smectic waveguides caused by dynamic distortions of the smectic layer planes are several orders of magnitude lower than in nematic waveguides [2,24], and SLCs may be used in nonlinear optical applications [2]. Recently, the reconfigurable smectic layer curvature has been studied [25]. The using of the external electric field to create the dynamic variations of the smectic layer configuration attracted a wide interest [25]. The different types of the periodic focal conic domain (FCD) arrays with the domain size, shape, orientation, and lattice symmetry controlled by external fields can be obtained [25]. The applications of SALC such as soft-lithographic templates, superhydrophobic surfaces, microlens arrays, and optically selective photomasks have been developed [25].

The nonlinear optical phenomena in SALC have been investigated theoretically [26–36]. It has been shown that the light self-focusing, self-trapping, Brillouin-like SLS, and four-wave mixing (FWM) related to the light enhanced smectic layer normal displacement $u(\vec{r},t)$ occur in SALC under certain conditions. The nonlinear effects based on this nonlinearity mechanism specific for SALC are strongly anisotropic, and the corresponding SLS gain coefficient is significantly larger than the one in the case of the Brillouin SLS in isotropic organic liquids. The nonlinear interaction of the surface plasmon polaritons (SPPs) in the metal-insulator-metal (MIM) waveguide has been analyzed [36]. In particular, it has been shown theoretically that the strong SLS of the transverse magnetic (TM) even modes can occur in the optical slab waveguide with a SALC core [35].

In this paper, we investigated in detail SLS in the Silicon-SALC slab waveguide. We discussed in detail the peculiarities of different types of LCs and concentrated on the optical properties of SALC. We derived the SALC layer equation of motion and the truncated equations for the optical wave slowly varying amplitudes (SVAs). We discussed the contribution of the TM even and odd modes and the transverse electric (TE) modes of the Silicon-SALC waveguide. We solved simultaneously the Maxwell equations including the nonlinear polarization for the waveguide modes and the equation of motion for the smectic layer normal displacement $u(\vec{r},t)$ in the optical wave field. We evaluated $u(\vec{r},t)$ and the hydrodynamic velocity $\vec{v}(\vec{r},t)$ in the SALC core of the waveguide. We obtained the novel explicit solutions for the SVAs of the interfering waveguide modes and made numerical estimations of the waveguide mode parameters and the gain. The results of the numerical estimations are presented in Figures 2–8. The paper is constructed as follows. The theoretical model is presented in Section 2. The nonlinear polarization in the waveguide SALC core is evaluated in Section 3. The SVAs of the pumping and signal TM waveguide modes and the hydrodynamic velocity of smectic layers are calculated in Section 4. The conclusions are presented in Section 5.

2. Theoretical Model

A typical LC slab waveguide represents a LC thin film with a thickness of about 1 µm sandwiched between two glass slides of lower refractive index than LC [2]. One of slides is covered with an organic film. The input laser radiation is inserted into the film via the coupling prism [2]. The laser excites the TE and/or TM modes in the film which are then introduced into the LC core [2]. Such a structure can be placed on a Si substrate [8]. One of the claddings can be made of SiO$_2$ [8]. For the sake of definiteness, we consider the homeotropically oriented SALC core where the molecular elongated axes are perpendicular to the waveguide claddings and the smectic layer planes parallel to them. The structure of the optical slab waveguide with a homeotropically oriented SALC core is shown in Figure 1.

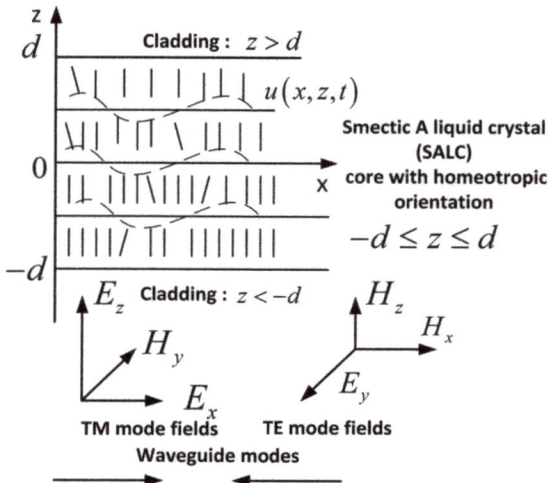

Figure 1. Optical slab waveguide with the homeotropically oriented SALC core of the thickness $2d$. $E_{x,z}$, H_y and E_y, $H_{x,z}$ are the electric and magnetic fields of the TM and TE modes, respectively.

Optical waves interact through the nonlinear polarization in a medium [37]. Generally, different types of SLS are described by the coupled wave equations for the light waves and for the corresponding material excitations [37]. The wave equation for electric field $\vec{E}(\vec{r}, t)$ of the optical wave propagating in a nonlinear medium has the form [37].

$$\text{curl curl } \vec{E} + \mu_0 \frac{\partial^2 \vec{D}^L}{\partial t^2} = -\mu_0 \frac{\partial^2 \vec{D}^{NL}}{\partial t^2} \qquad (1)$$

where μ_0 is the free space permeability, $\vec{D}^L$ and $\vec{D}^{NL}$ are the linear and nonlinear parts of the electric induction, respectively.

The SLS in the liquid crystalline waveguide with a SALC core is described by the coupled wave equations of the type (1) for the waveguide modes and the hydrodynamic equations for SALC. The SALC hydrodynamics in general case is very complicated taking into account the anisotropy and including the fluctuations of the mass density ρ, the layer displacement $u(\vec{r}, t)$ along the Z axis normal to the layers and the change of the director $\vec{n}$ [15,16]. The character of the fluctuation modes is determined by the propagation direction [15–18]. We assume that the SALC temperature is far from the temperature T_{SmA-N} of the SALC-NLC phase transition. Since the optical losses in SALC are negligible [2] the waveguide temperature is assumed to be constant and the smectic A phase is stable. In such a case, the system of hydrodynamic equations for SALC has the form [15].

$$\rho \frac{\partial v_i}{\partial t} = -\frac{\partial \Pi}{\partial x_i} + \Lambda_i + \frac{\partial \sigma'_{ik}}{\partial x_k} \qquad (2)$$

$$\Lambda_i = -\frac{\delta F}{\delta u_i} \qquad (3)$$

$$\sigma'_{ik} = \alpha_0 \delta_{ik} A_{ll} + \alpha_1 \delta_{iz} A_{zz} + \alpha_4 A_{ik} + \alpha_{56}(\delta_{iz} A_{zk} + \delta_{kz} A_{zi}) + \alpha_7 \delta_{iz} \delta_{kz} A_{ll} \qquad (4)$$

$$A_{ik} = \frac{1}{2}\left(\frac{\partial v_i}{\partial x_k} + \frac{\partial v_k}{\partial x_i}\right) \qquad (5)$$

$$\text{div } \vec{v} = 0 \qquad (6)$$

$$v_z = \frac{\partial u}{\partial t} \tag{7}$$

$$\delta_{ik} = \begin{cases} 1, & i = k \\ 0, & i \neq k \end{cases} \tag{8}$$

where $\vec{v}$ is the hydrodynamic velocity, Π is the pressure, $\vec{\Lambda}$ is the generalized force density, σ'_{ik} is the viscous stress tensor, α_i are the viscosity Leslie coefficients, F is the free energy density of SALC. The SALC free energy density in the presence of the external electric field $\vec{E}(\vec{r},t)$ has the form.

$$F = \frac{1}{2}B\left(\frac{\partial u}{\partial z}\right)^2 + \frac{1}{2}K\left(\frac{\partial^2 u}{\partial x^2} + \frac{\partial^2 u}{\partial y^2}\right)^2 - \frac{1}{2}\varepsilon_0 \varepsilon_{ik} E_i E_k \tag{9}$$

Here $B \sim 10^6 - 10^7$ J/m^3 is the elastic constant related to the layer compression, $K \sim 10^{-11}$ N is the Frank elastic constant associated with the SALC purely orientational energy, ε_0 is the free space permittivity, and ε_{ik} is the SALC permittivity tensor including the nonlinear terms related to the smectic layer strains. SALC is an optically uniaxial medium with the optical axis Z normal to the layer plane. It is given by [16].

$$\varepsilon_{xx} = \varepsilon_{yy} = \varepsilon_\perp + a_\perp \frac{\partial u}{\partial z} \tag{10}$$

$$\varepsilon_{zz} = \varepsilon_\| + a_\| \frac{\partial u}{\partial z} \tag{11}$$

$$\varepsilon_{xz} = \varepsilon_{zx} = -\varepsilon_a \frac{\partial u}{\partial x}, \varepsilon_{yz} = \varepsilon_{zy} = -\varepsilon_a \frac{\partial u}{\partial y} \tag{12}$$

where $\varepsilon_\perp, \varepsilon_\|$ are the diagonal components of the permittivity tensor perpendicular and parallel to the optical axis, respectively, $a_\perp \sim 1, a_\| \sim 1$ are the phenomenological dimensionless coefficients, and ε_a is the permittivity anisotropy. In our case, the losses in SALC can be neglected and the linear permittivity is real [2].

$$\varepsilon_a = \varepsilon_\| - \varepsilon_\perp \tag{13}$$

For the wave vector $\vec{k}_S$ oblique to the smectic layer plane in SALC there exist two practically uncoupled acoustic modes. One of these modes is the ordinary longitudinal sound wave caused by the mass density oscillations and described by the dispersion relation $\Omega = s_1 k_S$ independent of the propagation direction where the sound velocity $s_1 = \sqrt{A/\rho}$, and A is the elastic constant related to bulk compression [15–18]. The second mode is the so-called second sound (SS) with the following dispersion relation [15,17].

$$\Omega_{SS} = s_2 \frac{k_{S\perp} k_{Sz}}{k_S}, \quad s_2 = \sqrt{\frac{B}{\rho}} \tag{14}$$

where s_2 is the SS velocity, $k_{S\perp}, k_{Sz}$ are the SS wave vector components in the layer plane and normal to it, respectively. SS corresponds to the changes in the layer spacing, it is neither longitudinal, nor transverse, and vanishes for the wave vector parallel or perpendicular to the smectic layer plane as it is seen from Equation (14) [15–18]. Since the elastic constant $B \ll A \sim 10^9$ J/m^3, the SS may propagate in the SALC without the density change [15–18]. SS has been observed experimentally [19–21]. In such a case, SALC may be considered to be incompressible liquid according to Equation (6), the pressure $\Pi = 0$, and the SALC energy density F determined by Equation (9) does not include the bulk compression term. The purely orientational term second term in Equation (9) can be neglected since for the typical values of the elastic constants and $k_S \sim 10^5$ m^{-1} $B \gg Kk_S^2$. The normal layer displacement $u(\vec{r},t)$ by definition has only one component along the Z axis. Hence, the generalized force density $\vec{\Lambda}$ has only the z component according to Equation (3): $\vec{\Lambda} = (0,0,\Lambda_z)$. Equation (7) is specific for SALC since it determines the condition of the smectic layer continuity and the absence of the permeation process which can be neglected in the high frequency limit [15,17].

Taking into account the assumptions mentioned above and combining Equations (2)–(12) we obtain the equation of motion for smectic layer normal displacement $u\,(\vec{r},t)$ in an external electric field $\vec{E}\,(\vec{r},t)$ [36].

$$-\rho \nabla^2 \frac{\partial^2 u}{\partial t^2} + \left[\alpha_1 \nabla_\perp^2 \frac{\partial^2}{\partial z^2} + \frac{1}{2}(\alpha_4 + \alpha_{56}) \nabla^2 \nabla^2\right] \frac{\partial u}{\partial t} + B \nabla_\perp^2 \frac{\partial^2 u}{\partial z^2}$$
$$= \frac{\varepsilon_0}{2} \nabla_\perp^2 \left\{-2\varepsilon_a \left[\frac{\partial}{\partial x}(E_x E_z) + \frac{\partial}{\partial y}(E_y E_z)\right] + \frac{\partial}{\partial z}\left[a_\perp \left(E_x^2 + E_y^2\right) + a_\parallel E_z^2\right]\right\} \quad (15)$$

where $\nabla_\perp^2 = (\partial^2/\partial x^2) + (\partial^2/\partial y^2)$. Taking into account the SALC symmetry we may choose without the loss of generality the propagation plane in a slab waveguide as the xz plane. Then, using expressions (10)–(12) we obtain for the linear and nonlinear parts of the electric induction $\vec{D}^L$ and $\vec{D}^{NL}$.

$$D^L_{x,y} = \varepsilon_0 \varepsilon_\perp E_{x,y}, \, D^L_z = \varepsilon_0 \varepsilon_\parallel E_z \quad (16)$$

$$D^{NL}_x = \varepsilon_0 \left(a_\perp \frac{\partial u}{\partial z} E_x - \varepsilon_a \frac{\partial u}{\partial x} E_z\right); D^{NL}_y = \varepsilon_0 a_\perp \frac{\partial u}{\partial z} E_y \quad (17)$$

$$D^{NL}_z = \varepsilon_0 \left(a_\parallel \frac{\partial u}{\partial z} E_z - \varepsilon_a \frac{\partial u}{\partial x} E_x\right) \quad (18)$$

It is seen from Equations (17) and (18) that the nonlinear polarization in SALC is related to the smectic layer normal and tangential strain $\partial u/\partial z$ and $\partial u/\partial x$ as it was mentioned above [26–36]. We solve the wave Equation (1) according to the SVA approximation procedure [37]. In the linear approximation, we solve the homogeneous part of Equation (1) neglecting the nonlinear polarization (17) and (18).

$$\text{curl curl}\,\vec{E} + \mu_0 \frac{\partial^2 \vec{D}^L}{\partial t^2} = 0 \quad (19)$$

We obtain from Equation (19) the general solution and the linear dispersion relations for the waveguide modes [38,39]. Then, we evaluate the nonlinear polarization (17) and (18), derive the truncated equations for the SVAs of the waveguide mode electric fields in the SALC core and evaluate the complex SVA magnitudes and phases [37,38]. In the next section, we evaluate the waveguide modes and the nonlinear polarization defined by Equations (17) and (18).

3. Nonlinear Polarization in the SALC Core of the Waveguide

The TM and TE mode electric and magnetic fields have the form, respectively [38–40].

$$\vec{H}_{TM}(x,z,t) = H_y(x,z,t)\,\mathbf{a}_y; \, \vec{E}_{TM}(x,z,t) = (E_x(x,z,t), 0, E_z(x,z,t)) \quad (20)$$

$$\vec{E}_{TE}(x,z,t) = E_y(x,z,t)\,\mathbf{a}_y; \, \vec{H}_{TE}(x,z,t) = (H_x(x,z,t), 0, H_z(x,z,t)) \quad (21)$$

We consider separately the TM and TE modes propagating in the slab optical waveguide with the SALC core because Equations (15)–(18) show that in the framework of the slab waveguide model TE and TM modes do not interact. We start with the analysis of the TM even modes. Assuming that the waveguide is symmetric with the identical claddings $z > d$, $z < -d$ characterized by the same permittivity ε_{r2} and the refraction index $n_2 = \sqrt{\varepsilon_{r2}}$, solving Equation (1) in the linear approximation and using the boundary conditions for the tangential components of the magnetic and electric field in the cladding H_{yC} and E_{xC} and in the SALC core H_{ySA} and E_{xSA}, respectively [38–40].

$$H_{yC}(z=d) = H_{ySA}(z=d); E_{xC}(z=d) = E_{xSA}(z=d) \quad (22)$$

we obtain for the electric field $E_{x,zSA}$, $E_{x,zC}$ in the SALC core $|z| \leq d$ and in the cladding $z > d, z < -d$, respectively [35,39].

$$E_{xSA} = -iE_{0zSA} \frac{k\varepsilon_\parallel}{\beta\varepsilon_\perp} \sin kz \exp\left[i\left(\omega t - \beta x\right)\right] \tag{23}$$

$$E_{zSA} = -E_{0zSA} \cos kz \exp\left[i\left(\omega t - \beta x\right)\right] \tag{24}$$

$$E_{xC} = \begin{cases} i\frac{\alpha}{\beta} E_{0zC} \exp(-\alpha z) \exp i\left(\omega t - \beta x\right), & z > d \\ -i\frac{\alpha}{\beta} E_{0zC} \exp(\alpha z) \exp i\left(\omega t - \beta x\right), & z < -d \end{cases} \tag{25}$$

$$E_{zC} = \begin{cases} E_{0zC} \exp(-\alpha z) \exp i\left(\omega t - \beta x\right), & z > d \\ E_{0zC} \exp(\alpha z) \exp i\left(\omega t - \beta x\right), & z < -d \end{cases} \tag{26}$$

Here ω is the optical mode angular frequency, β is the propagation constant, k is the wave vector in the core, and α is the wavenumber in the cladding. They are given by

$$\beta = \sqrt{\varepsilon_\parallel \left[\left(\frac{\omega}{c}\right)^2 - \frac{k^2}{\varepsilon_\perp}\right]} \tag{27}$$

$$\alpha = \sqrt{\beta^2 - \frac{\omega^2}{c^2} \varepsilon_{r2}} \tag{28}$$

Expression (27) shows that the TM mode propagates in an anisotropic medium as an extraordinary wave [41]. The wave vector k for the TM even modes is defined by the dispersion relation

$$\tan kd = \frac{\varepsilon_\perp}{\varepsilon_{r2}} \sqrt{\left(\frac{V}{kd}\right)^2 - \frac{\varepsilon_\parallel}{\varepsilon_\perp}}; V = \frac{2\pi d}{\lambda_0} \sqrt{\varepsilon_\parallel - \varepsilon_{r2}}; \varepsilon_\parallel > \varepsilon_{r2} \tag{29}$$

where $\lambda_0 = 2\pi c/\omega$ and c are the free space wavelength and light velocity, respectively. Consider now the TM odd modes. In this case, the electric field components $E^{odd}_{x,zSA}$ in the SALC has the form [39].

$$E^{odd}_{zSA} = E^{odd}_{0zSA} \sin kz \exp i\left(\omega t - \beta x\right) \tag{30}$$

$$E^{odd}_{xSA} = -\frac{ik\varepsilon_\parallel}{\beta\varepsilon_\perp} E^{odd}_{0zSA} \cos kz \exp i\left(\omega t - \beta x\right) \tag{31}$$

The boundary conditions (22) give the following dispersion relation for the TM odd modes.

$$-\cot kd = \frac{\varepsilon_\perp}{\varepsilon_{r2}} \sqrt{\left(\frac{V}{kd}\right)^2 - \frac{\varepsilon_\parallel}{\varepsilon_\perp}} \tag{32}$$

The solution of the dispersion relations (29) and (32) for the TM even and odd modes are presented in Figure 2a,b, respectively. It is seen from Figure 2a,b that for the frequency $\omega = 5\pi \times 10^{14}$ s^{-1} and for the typical values of the waveguide parameters there exist two even TM modes TM$^{even}_{0,1}$ and one odd TM mode TM$^{odd}_1$. The normalized wavenumber kd and propagation constant βd dependence on the optical wavelength λ for the even modes TM$^{even}_{0,1}$ and for the odd mode TM$^{odd}_1$ are presented in Figure 3a,b, respectively. The normalized wavenumber in the cladding αd spectral dependence is shown in Figure 4. It is seen from Figure 4 that the fundamental even mode TM$^{even}_0$ does not have a cutoff while the second even mode TM$^{even}_1$ has a cutoff wavelength coinciding with the cutoff wavelength in Figure 3a,b, respectively. Comparison of Figures 3a and 4 shows that in the wavelength region under consideration $kd > \pi/2$, and $\alpha d \neq 0$ for the odd mode TM$^{odd}_1$ [39]. The solutions of the dispersion relations (29) and (32) presented in Figure 3a,b show that for the waveguide SALC core thickness of $2d = 2$ μm, the typical values of LC and cladding permittivity [8], and the wavelengths $\lambda_0 \approx$ 1.4–1.55 μm important for optical communications the single mode regime occurs. We consider the interaction of the TM modes with the close optical frequencies $\omega_{1,2}$ such that the frequency shift $\Delta\omega = \omega_1 - \omega_2 \sim 10^8$–$10^9$ s$^{-1} \ll \omega_1$ which is typical for the light scattering in SALC [15,19]. The numerical estimations of the propagation constant β and the wave vector k according to Equations (27) and (29) show that for the frequency shifts $\Delta\omega \sim 10^8$–10^9 s^{-1} the values of β and k are practically the same for the TM modes with the close frequencies $\omega_{1,2}$. Consequently, the strong interaction occurs only for the counter-propagating TM modes. For the sake of definiteness, we consider the interaction of the TM even modes (23) and (24). Obviously, the nonlinear interaction of the TM odd modes would be practically the same. Using expressions (23) and (24) we can write for such TM even mode electric field [35].

$$\vec{E}_{SA1,2} = \frac{1}{2} E_{0zSA1,2}(x,t) \left\{ -\vec{a}_x i \frac{k\varepsilon_\parallel}{\beta\varepsilon_\perp} \sin kz \mp \vec{a}_z \cos kz \right\} \exp\left[i(\omega_{1,2} t \mp \beta x)\right] + c.c. \quad (33)$$

where c.c. stands for complex conjugate, and $\vec{a}_x$, $\vec{a}_z$ are the unit vectors of the X and Z axes, respectively. We assume that $E_{0zSA1,2}(x,t) = |E_{0zSA1,2}| \exp i\theta_{1,2}$ are the complex SVAs [37].

$$\left|\frac{\partial^2 E_{0zSA1,2}}{\partial x^2}\right| \ll \left|\beta\frac{\partial E_{0zSA1,2}}{\partial x}\right| ; \left|\frac{\partial^2 E_{0zSA1,2}}{\partial t^2}\right| \ll \left|\omega\frac{\partial E_{0zSA1,2}}{\partial t}\right| \quad (34)$$

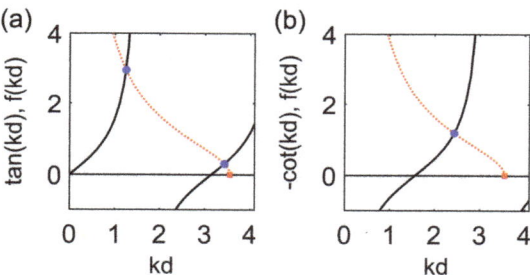

Figure 2. Graphic solution of the disperison relations for the TM even modes (**a**) and odd modes (**b**); $f(kd) = \frac{\varepsilon_\perp}{\varepsilon_{r2}} \sqrt{\left(\frac{V}{kd}\right)^2 - \frac{\varepsilon_\parallel}{\varepsilon_\perp}}$, $\omega = 5\pi \times 10^{14} \mathrm{s}^{-1}$.

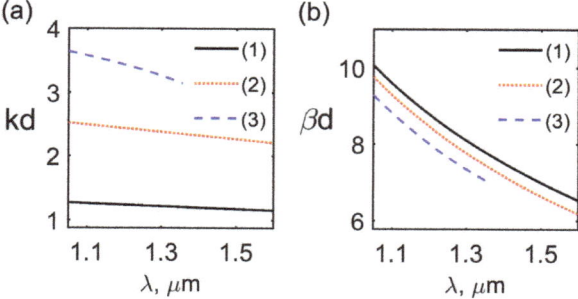

Figure 3. The normalized wavenumber kd (**a**) and propagation constant βd (**b**) dependence on the optical wavelength λ for the even modes TM$_{0,1}^{even}$ (curves 1, 3) and the odd mode TM$_1^{odd}$ (curve 2).

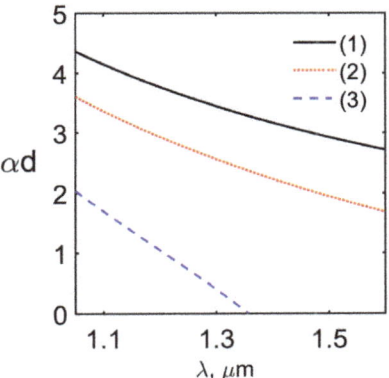

Figure 4. The dependence of normalized wavenumber in the cladding αd for the even modes $TM_{0,1}^{even}$ (curves 1, 3) and odd mode TM_1^{odd} (curve 2) on the optical wavelength.

At the small distances of several mm typical for the optical waveguide length the dependence of SVAs on x and the dispersion effects can be neglected, and the SVAs $E_{0zSA1,2}(t)$ depend only on time. Substituting expressions (33) into equation of motion (15) and keeping in the right-hand side (RHS) only the terms with the frequency difference $\Delta \omega$ we obtain.

$$-\rho \nabla^2 \frac{\partial^2 u}{\partial t^2} + \left[\alpha_1 \frac{\partial^2}{\partial x^2} \frac{\partial^2}{\partial z^2} + \frac{1}{2}(\alpha_4 + \alpha_{56}) \nabla^2 \nabla^2 \right] \frac{\partial u}{\partial t} + B \frac{\partial^2}{\partial x^2} \frac{\partial^2 u}{\partial z^2}$$
$$= -2\varepsilon_0 \beta^2 k E_{0zSA1} E^*_{0zSA2} \left[\varepsilon_a \frac{\varepsilon_\parallel}{\varepsilon_\perp} + a_\perp \frac{1}{2} \left(\frac{k\varepsilon_\parallel}{\beta \varepsilon_\perp} \right)^2 + \frac{1}{2} a_\parallel \right] \quad (35)$$
$$\times \sin 2kz \exp\{i[(\omega_1 - \omega_2)t - 2\beta x]\} + c.c.$$

Then the particular solution of Equation (35) related to its RHS yields the dynamic grating of the smectic layer normal displacement.

$$u(x,z,t) = U_0 \sin 2kz \exp\{i[(\omega_1 - \omega_2)t - 2\beta x]\} + c.c. \quad (36)$$

where

$$U_0 = \frac{\varepsilon_0 \beta^2 k E_{0zSA1} E^*_{0zSA2} \left[\varepsilon_a \frac{\varepsilon_\parallel}{\varepsilon_\perp} + a_\perp \frac{1}{2} \left(\frac{k\varepsilon_\parallel}{\beta \varepsilon_\perp} \right)^2 + \frac{1}{2} a_\parallel \right]}{2\rho(\beta^2 + k^2) G(k, \beta, \Delta\omega)} \quad (37)$$

$$G(k, \beta, \Delta\omega) = (\Delta\omega)^2 - i\Delta\omega \Gamma - \Omega^2 \quad (38)$$

$$\Gamma = \frac{1}{\rho} \left[4 \frac{\alpha_1 \beta^2 k^2}{(\beta^2 + k^2)} + 2(\alpha_4 + \alpha_{56})(\beta^2 + k^2) \right]; \quad \Omega^2 = 4 \frac{B\beta^2 k^2}{\rho(\beta^2 + k^2)} \quad (39)$$

Here Ω, Γ are SS frequency and decay factor, respectively [15–21]. The SS frequency Ω and decay factor Γ dependence on the optical wavelength λ for the first two TM modes are presented in Figure 5a,b, respectively. Numerical estimations show that for the typical values of SALC parameters [15–21], the optical wavelength in the range of $\lambda_{opt} \sim$ 1.3–1.55 μm and $\Delta\omega \sim 10^8$–10^9 s^{-1} the homogeneous layer oscillations are overdamped. For this reason, the rapidly decaying homogeneous solution of Equation (35) can be neglected. We have taken into account only the solution (36) enhanced by the interfering optical TM modes (33).

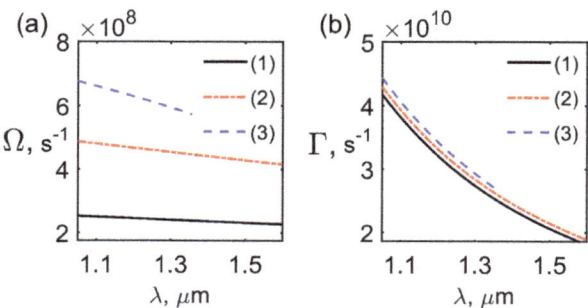

Figure 5. The SS frequency Ω (a) and decay constant Γ (b) dependence on the optical wavelength for the even modes $TM_{0,1}^{even}$ (curves 1,3) and the odd mode TM_1^{odd} (curve 2).

Substituting expressions (33) and (36) into Equations (17) and (18) we evaluate the nonlinear part of the electric induction $\vec{D}^{NL} = (D_x^{NL}, 0, D_z^{NL})$ which has only x and z components for the TM modes. Separating the phase matched parts of $\vec{D}^{NL}$ with the frequencies $\omega_{1,2}$, respectively, we obtain.

$$D_x^{NL}(\omega_1) = \varepsilon_0 U_0 i\beta E_{0zSA2} \sin kz \exp\left[i(\omega_1 t - \beta x)\right]$$
$$\times \left\{ -a_\perp \frac{k^2 \varepsilon_\parallel}{\beta^2 \varepsilon_\perp} \cos 2kz + 2\varepsilon_a \cos^2 kz \right\} \quad (40)$$

$$D_z^{NL}(\omega_1) = \varepsilon_0 k U_0 E_{0zSA2} \cos kz \exp\left[i(\omega_1 t - \beta x)\right]$$
$$\times \left\{ a_\parallel \cos 2kz + 2\varepsilon_a \frac{\varepsilon_\parallel}{\varepsilon_\perp} \sin^2 kz \right\} \quad (41)$$

$$D_x^{NL}(\omega_2) = \varepsilon_0 U_0^* i E_{0zSA1} \sin kz \exp\left[i(\omega_2 t + \beta x)\right]$$
$$\times \left\{ -a_\perp \frac{k^2 \varepsilon_\parallel}{\beta \varepsilon_\perp} \cos 2kz + 2\varepsilon_a \beta \cos^2 kz \right\} \quad (42)$$

$$D_z^{NL}(\omega_2) = -\varepsilon_0 k U_0^* E_{0zSA1} \cos kz \exp\left[i(\omega_2 t + \beta x)\right]$$
$$\times \left\{ a_\parallel \cos 2kz + 2\varepsilon_a \frac{\varepsilon_\parallel}{\varepsilon_\perp} \sin^2 kz \right\} \quad (43)$$

The nonlinear polarization (40)–(43) is related to the specific cubic nonlinearity related to the smectic layer displacement which occurs without the change of the SALC mass density.

The electric field $E_{0ySA1,2}$ of the TE modes (21) is perpendicular to the optical axis Z. It has the form.

$$\vec{E}_{SA1,2} = \frac{1}{2} E_{0ySA1,2}(x,t) \vec{a}_y \cos kz \exp\left[i(\omega_{1,2}t \mp \beta x)\right] + c.c.$$

TE modes propagate in an anisotropic medium as ordinary waves with the propagation constant $\beta^2 = (\omega^2 \varepsilon_\perp / c^2) - k^2$ including only the transverse permittivity $\varepsilon_\perp$ [41]. The boundary conditions for the TE modes have the form.

$$E_{yC}(d) = E_{ySA}(d); H_{zC}(d) = H_{zSA}(d) \quad (44)$$

They yield the TE mode dispersion relation similar to the isotropic medium [38].

$$\tan kd = \sqrt{\frac{V_{TE}^2}{(kd)^2} - 1}; \quad V = \frac{2\pi d}{\lambda_0}\sqrt{\varepsilon_\perp - \varepsilon_{r2}}, \quad \varepsilon_\perp > \varepsilon_{r2} \quad (45)$$

In LC typically $\varepsilon_\parallel > \varepsilon_\perp$ [2,14], and under the condition $\varepsilon_\parallel > \varepsilon_{r2} > \varepsilon_\perp$ only TM modes can propagate in the slab optical waveguide. In general case, the nonlinear polarization enhanced by the TE modes includes only the component $D_y^{NL} = \varepsilon_0 a_\perp (\partial u/\partial z) E_y$ as it is seen from expression (17), and the dynamic grating amplitude $U_{0TE} \sim E_{0ySA1} E_{0ySA2}^*$. Obviously, the nonlinear interaction of the TE modes is isotropic and less pronounced than the TM mode interaction including both the longitudinal and the transverse component of the electric field. For this reason, we analyze in detail the TM mode nonlinear interaction.

4. Evaluation of the TM Mode SVAs

Using the standard procedure [37], we substitute expressions (33), (16), and (40)–(43) into Equation (1), separate the linear and nonlinear parts, neglect the small terms $\sim |\partial^2 E_{0zSA1,2}/\partial t^2|$ according to the SVA approximation condition (34) and equate the phase matched terms the frequencies $\omega_{1,2}$, respectively. Then we obtain the coupled equations for the SVAs $E_{0zSA1,2}(t)$.

$$\varepsilon_\parallel \frac{\partial E_{0zSA1}}{\partial t} \left\{ \mathbf{a}_x \frac{k}{\beta} \sin kz - \mathbf{a}_z i \cos kz \right\}$$
$$= \omega_1 U_0 E_{0zSA2} \left\{ \mathbf{a}_x i\beta \sin kz \left[-a_\perp \frac{k^2 \varepsilon_\parallel}{\beta^2 \varepsilon_\perp} \cos 2kz + 2\varepsilon_a \cos^2 kz \right] \right. \tag{46}$$
$$\left. + \mathbf{a}_z k \cos kz \left[a_\parallel \cos 2kz + 2\varepsilon_a \frac{\varepsilon_\parallel}{\varepsilon_\perp} \sin^2 kz \right] \right\}$$

$$\varepsilon_\parallel \frac{\partial E_{0zSA2}}{\partial t} \left\{ \mathbf{a}_x \frac{k}{\beta} \sin kz + \mathbf{a}_z i \cos kz \right\}$$
$$= \omega_2 U_0^* E_{0zSA1} \left\{ \mathbf{a}_x i\beta \sin kz \left[-a_\perp \frac{k^2 \varepsilon_\parallel}{\beta^2 \varepsilon_\perp} \cos 2kz + 2\varepsilon_a \cos^2 kz \right] \right. \tag{47}$$
$$\left. - \mathbf{a}_z k \cos kz \left[a_\parallel \cos 2kz + 2\varepsilon_a \frac{\varepsilon_\parallel}{\varepsilon_\perp} \sin^2 kz \right] \right\}$$

We multiply Equations (46) and (47) by the vectors $\left\{ \mathbf{a}_x \frac{k}{\beta} \sin kz - \mathbf{a}_z i \cos kz \right\}^*$ and $\left\{ \mathbf{a}_x \frac{k}{\beta} \sin kz + \mathbf{a}_z i \cos kz \right\}^*$, respectively, substitute the SVA expressions

$$E_{0zSA1,2}(x,t) = |E_{0zSA1,2}| \exp i\theta_{1,2} \tag{48}$$

and separate the real and imaginary parts of the resulting equations. Then we obtain the following equations for the magnitudes $|E_{0zSA1,2}|$ and phases $\theta_{1,2}$ of the TM mode SVAs.

$$\frac{1}{\omega_1} \frac{\partial |E_{0zSA1}|^2}{\partial t} F_1(z) = \frac{\varepsilon_0 |E_{0zSA1}|^2 |E_{0zSA2}|^2 \beta^2 k^2 h \operatorname{Im} G(k,\beta,\Delta\omega)}{\varepsilon_\parallel \rho (\beta^2 + k^2) |G(k,\beta,\Delta\omega)|^2} F_2(z) \tag{49}$$

$$\frac{1}{\omega_2} \frac{\partial |E_{0zSA2}|^2}{\partial t} F_1(z) = -\frac{\varepsilon_0 |E_{0zSA1}|^2 |E_{0zSA2}|^2 \beta^2 k^2 h \operatorname{Im} G(k,\beta,\Delta\omega)}{\varepsilon_\parallel \rho (\beta^2 + k^2) |G(k,\beta,\Delta\omega)|^2} F_2(z) \tag{50}$$

$$\varepsilon_\parallel \frac{\partial \theta_1}{\partial t} F_1(z) = \omega_1 \varepsilon_0 |E_{0zSA2}|^2 \frac{\beta^2 k^2 h \operatorname{Re} G(k,\beta,\Delta\omega)}{2\rho (\beta^2 + k^2) |G(k,\beta,\Delta\omega)|^2} F_2(z) \tag{51}$$

$$\varepsilon_\parallel \frac{\partial \theta_2}{\partial t} F_1(z) = \omega_2 \varepsilon_0 \frac{\beta^2 k^2 |E_{0zSA1}|^2 h \operatorname{Re} G(k,\beta,\Delta\omega)}{2\rho (\beta^2 + k^2) |G(k,\beta,\Delta\omega)|^2} F_2(z) \tag{52}$$

where

$$F_1(z) = \left(\frac{k}{\beta}\right)^2 \sin^2 kz + \cos^2 kz; \quad F_2(z) = \{\sin^2 kz \left[-a_\perp \frac{k^2 \varepsilon_\parallel}{\beta^2 \varepsilon_\perp} \cos 2kz + 2\varepsilon_a \cos^2 kz\right] \\ + \cos^2 kz \left[a_\parallel \cos 2kz + 2\varepsilon_a \frac{\varepsilon_\parallel}{\varepsilon_\perp} \sin^2 kz\right]\} \tag{53}$$

and

$$h = \varepsilon_a \frac{\varepsilon_\parallel}{\varepsilon_\perp} + a_\perp \frac{1}{2} \left(\frac{k \varepsilon_\parallel}{\beta \varepsilon_\perp}\right)^2 + \frac{1}{2} a_\parallel \tag{54}$$

Combining Equations (49) and (50) we obtain for the SVA magnitudes $|E_{0zSA1,2}|$

$$\frac{\partial}{\partial t}\left(\frac{|E_{0zSA1}|^2}{\omega_1} + \frac{|E_{0zSA2}|^2}{\omega_2}\right) = 0 \tag{55}$$

and

$$\frac{|E_{0zSA1}|^2}{\omega_1} + \frac{|E_{0zSA2}|^2}{\omega_2} = const = I_0 \tag{56}$$

where

$$I_0 = \frac{|E_{0zSA1}(0)|^2}{\omega_1} + \frac{|E_{0zSA2}(0)|^2}{\omega_2} \tag{57}$$

Equation (56) is the Manley-Rowe relation for the SVA magnitudes $|E_{0zSA1,2}|$ which corresponds to the conservation of the photon number in the SLS process [37]. It is seen from Equation (38) that for $\Delta\omega = \omega_1 - \omega_2 > 0$ the imaginary part $\mathrm{Im}G(k, \beta, \Delta\omega) < 0$, and the intensity $|E_{0zSA1}|^2$ of the TM mode with the higher frequency ω_1 is decreasing with time while the intensity $|E_{0zSA2}|^2$ of the TM mode with the lower frequency ω_2 is increasing. Consequently, the TM modes with the frequencies $\omega_{1,2}$ are the pumping and signal waves, respectively, and the Stokes type SLS occurs [37]. Equations (49) and (50) describe the energy exchange between the TM modes, while Equations (51) and (52) describe the cross-phase modulation (XPM) process.

We introduce the dimensionless variables

$$I_{1,2} = \frac{|E_{0zSA1,2}|^2}{\omega_{1,2} I_0} \tag{58}$$

such that $I_1 + I_2 = 1$. Substituting expressions (58) into Equations (49) and (50), integrating both parts of these equations over z from $-d$ up to d and using the Manley-Rowe relation (56) we obtain the following solutions for the normalized SVA intensities $I_{1,2}$.

$$I_1(t) = \frac{I_1(0)}{\{(1 - I_1(0))\exp[gF(kd)t] + I_1(0)\}} \tag{59}$$

$$I_2(t) = \frac{(1 - I_1(0))}{(1 - I_1(0)) + I_1(0)\exp[-gF(kd)t]} \tag{60}$$

where the gain g and the geometric factor $F(kd)$ are given by.

$$g = \frac{\varepsilon_0}{\varepsilon_\parallel} \frac{\omega_1 \omega_2 I_0 \beta^2 k^2 h |\mathrm{Im}G(k,\beta,\Delta\omega)|}{\rho(\beta^2 + k^2)|G(k,\beta,\Delta\omega)|^2} > 0 \tag{61}$$

$$F(kd) = \{\left[a_\perp \frac{k^2 \varepsilon_\|}{\beta^2 \varepsilon_\perp} + \varepsilon_a \left(1 + \frac{\varepsilon_\|}{\varepsilon_\perp}\right) + a_\|\right] kd + \left[-a_\perp \frac{k^2 \varepsilon_\|}{\beta^2 \varepsilon_\perp} + a_\|\right] \sin 2kd$$

$$+ \frac{1}{4}\left[a_\perp \frac{k^2 \varepsilon_\|}{\beta^2 \varepsilon_\perp} - \varepsilon_a \left(1 + \frac{\varepsilon_\|}{\varepsilon_\perp}\right) + a_\|\right] \sin 4kd\} \left\{\left[\left(\frac{k}{\beta}\right)^2 + 1\right] 2kd + \left[1 - \left(\frac{k}{\beta}\right)^2\right] \sin 2kd\right\}^{-1} \quad (62)$$

The pumping intensity threshold in the SLS process described by expressions (59) and (60) is absent since the losses in SALC can be neglected as it was mentioned above. Comparison of expressions (59) and (60) shows that for the initial pumping wave intensity larger than the initial signal wave intensity $I_1(0) > I_2(0)$ the crossing time $t_0 > 0$ exists where $I_1(t_0) = I_2(t_0)$. It is given by.

$$t_0 = \frac{1}{gF(kd)} \ln\left[\frac{I_1(0)}{I_2(0)}\right] \quad (63)$$

Substituting expression (63) into expressions (59) and (60) we obtain.

$$I_{1,2}(t) = \frac{1}{2}\left\{1 \mp \tanh\left[\frac{1}{2}gF(kd)(t - t_0)\right]\right\} \quad (64)$$

The spectral dependence of the gain g and its dependence on the normalized intensity I_0 are presented in Figure 6a,b.

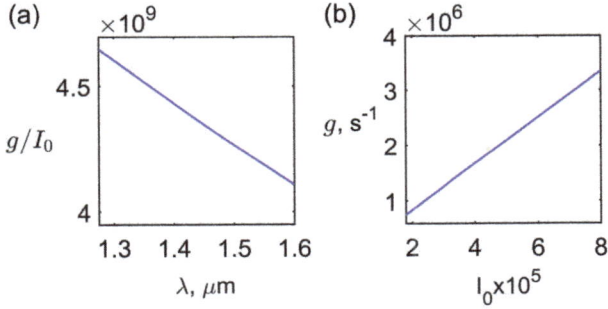

Figure 6. The normalized gain g/I_0 ($s^{-2}V^{-2}m^2$) dependence on the optical wavelength λ (**a**); the gain g dependence on the normalized intensity I_0 ($V^2 m^{-2} s$) for the optical wavelength $\lambda = 1.55$ µm (**b**).

Figure 6a shows that the gain is slightly varying in the optical wavelength range of interest because $\Gamma \gg \Omega$ as it is seen from Figure 5a,b. The gain g has a maximum value g_{max} at the SS resonance condition when $\Delta\omega = \Omega$ and $\text{Re}G(k, \beta, \Delta\omega) = 0$ according to expression (38). The numerical estimations show that for the typical values of $k, \beta \sim 10^6$ m^{-1} and $\Delta\omega \sim 10^8 - 10^9$ s^{-1} the SS resonance condition can be satisfied. The numerical estimations also show that for the values of kd defined by the dispersion relation (29) $F(kd) \sim 1$. The dependence of the gain g on the normalized intensity I_0 is linear as it is seen from Figure 6b. Such a dependence is typical for the Brillouin and Rayleigh SLS [37]. The SLS in our case is essentially orientational since the optical nonlinearity mechanism is related to the SALC layer displacement and occurs without the mass density change [15]. For the feasible optical wave electric fields E the condition $\varepsilon_0 E^2/B \ll 1$ is always valid, and the gain saturation does not take place.

It is seen from expressions (59) and (60) that for $t \to \infty$ the pumping wave intensity is depleted $I_1(t) \to 0$ while the signal wave intensity is amplified up to the saturation level $I_2(t) \to 1$. The time dependence for the normalized intensities $I_{1,2}(t)$ for the initial conditions $I_1(0) = 0.8$, $I_2(0) = 0.2$, pumping wavelength $\lambda_1 = 1.55$ µm and the pumping wave electric field amplitude $E_{0zSA1} = 10^5$ V/m, 10^6 V/m is shown in Figure 7. It is seen from Figure 7a,b that the amplified signal wave rise time is about 60 µsec and 0.6 µsec for the feasible electric field $\sim 10^5$ V/m, 10^6 V/m, respectively, which is much faster than the director axis relaxation time $\tau_r \sim 1$ ms in NLC [14].

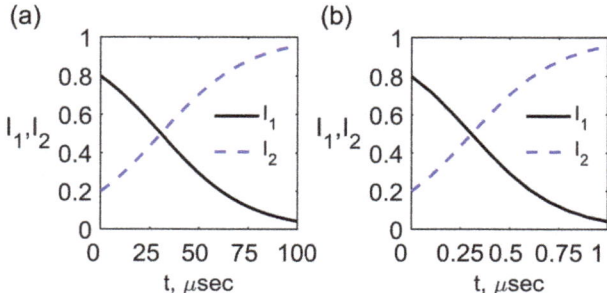

Figure 7. The time dependence of the normalized intensities $I_{1,2}(t)$ for the initial conditions $I_1(0) = 0.8$, $I_2(0) = 0.2$, pumping wavelength $\lambda_1 = 1.55$ µm and the pumping wave electric field amplitude $E_{0zSA1} = 10^5$ V/m (**a**) and 10^6 V/m (**b**).

Integrating both parts of Equations (51) and (52) over z from $-d$ up to d and substituting expressions (58)–(60), (62) into these equations, we obtain the expressions for the pumping and signal wave phases $\theta_{1,2}$. They have the form.

$$\theta_1(t) = \frac{\text{Re}G(k,\beta,\Delta\omega)}{2|\text{Im}G(k,\beta,\Delta\omega)|} \ln\{\exp[gF(kd)t](1 - I_1(0)) + I_1(0)\} \tag{65}$$

$$\theta_2(t) = -\frac{\text{Re}G(k,\beta,\Delta\omega)}{2|\text{Im}G(k,\beta,\Delta\omega)|} \ln\{I_1(0)\exp[-gF(kd)t] + (1 - I_1(0))\} \tag{66}$$

It is seen from Equations (65) and (66) that XPM occurs, and the depletion of the pumping wave is accompanied by rapid linear increase of its phase $\theta_1(t)$ which corresponds to the fast oscillations of the amplitude E_{0zSA1}.

$$t \to \infty, \; \theta_1(t) \approx \frac{\text{Re}G(k,\beta,\Delta\omega)}{2|\text{Im}G(k,\beta,\Delta\omega)|}(gF(kd)t) \to \infty \tag{67}$$

The phase of the amplified signal wave $\theta_2(t)$ tends to the constant level:

$$t \to \infty, \; \theta_2(t) \to -\frac{\text{Re}G(k,\beta,\Delta\omega)}{2|\text{Im}G(k,\beta,\Delta\omega)|} \ln I_2(0) \tag{68}$$

The temporal evolution of $\cos\theta_{1,2}(t)$ is shown in Figure 8a,b, respectively. The characteristic time of the phase variation is about 10^{-4} s for the pumping wave electric field amplitude $E_{0zSA1} = 10^5$ V/m. The comparison of expressions (38), (61), (65) and (66) shows that in the SS resonance case $\Delta\omega = \Omega$, $\text{Re}G(k,\beta,\Delta\omega) = 0$, XPM is absent: $\theta_{1,2} = const$.

Consider now the hydrodynamic behavior of the SALC core. Substituting expressions (48), (58) and (64) into expression (37) we obtain the explicit expression of the smectic layer grating amplitude U_0.

$$U_0 = \frac{\varepsilon_0 \beta^2 k I_0 \sqrt{\omega_1\omega_2}\left[\varepsilon_a \frac{\varepsilon_\parallel}{\varepsilon_\perp} + a_\perp \frac{1}{2}\left(\frac{k\varepsilon_\parallel}{\beta\varepsilon_\perp}\right)^2 + \frac{1}{2}a_\parallel\right]}{4\rho(\beta^2 + k^2)G(k,\beta,\Delta\omega)} \frac{\exp i(\theta_1 - \theta_2)}{\cosh\left[\frac{1}{2}gF(kd)(t-t_0)\right]} \tag{69}$$

It is seen from Equation (69) that the crossing time t_0 corresponds to the maximum of the smectic layer strain pulse. Substituting expressions (36) and (69) into Equations (6) and (7) we obtain the following expressions of the hydrodynamic velocity components $v_{x,z}(x,z,t)$.

$$v_x = \frac{k\Delta\omega}{\beta} U_0 \cos 2kz \exp i\left[(\omega_1 - \omega_2)t - 2\beta x\right] + c.c. \qquad (70)$$

$$v_z = i\Delta\omega U_0 \sin 2kz \exp i\left[(\omega_1 - \omega_2)t - 2\beta x\right] + c.c. \qquad (71)$$

It is seen from expressions (70) and (71) that they also have the form of the pulses (69).

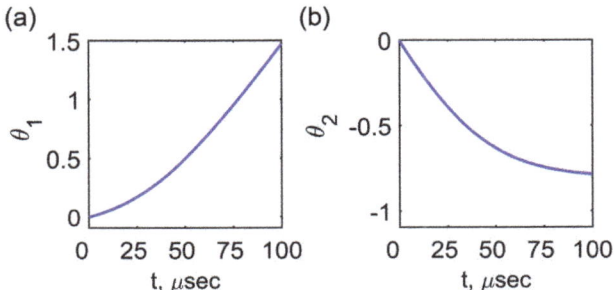

Figure 8. The temporal evolution of the phases $\theta_1(t)$ (**a**) and $\theta_2(t)$ (**b**) for the pumping wave electric field amplitude $E_{0zSA1} = 10^5$ V/m and the pumping wavelength $\lambda_1 = 1.33$ μm.

5. Conclusions

We investigated theoretically the nonlinear optical phenomena in the optical slab waveguide with the SALC core. We calculated the TM and TE modes in such a strongly anisotropic waveguide. We have shown that the single mode regime can be realized for the waveguide core thickness of about 1–2 μm and optical wavelength of $\lambda_{opt} \sim$ 1.35–1.55 μm important for the optical communication applications. The cubic nonlinearity of SALC is related to the smectic layer normal displacement. The nonlinear interaction is especially strong for the counter-propagating TM modes. We solved simultaneously the equation of motion for the smectic layer normal displacement in the optical field and the wave equation for the TM mode electric field using SVA approximation. The interfering optical fields create the smectic layer displacement dynamic grating which propagates in SALC without the mass density change. As a result the nonlinear polarization occurs and the SLS accompanied by XPM takes place in the waveguide. We evaluated the pumping and signal TM mode SVA magnitudes and phases. In the resonance case when the TM mode frequency difference $\Delta\omega$ equals to the SS frequency Ω the gain g has a maximum value, and XPM is absent. The smectic layer strain has a pulse form with a maximum corresponding to the crossing time of the pumping and signal TM modes. We also evaluated the hydrodynamic velocity enhanced by the interfering TM modes. The numerical estimations show that the SLS in SALC is much faster than the light scattering in NLC related to the director reorientation.

Author Contributions: Conceptualization, B.I.L.; Methodology, B.I.L. and Y.B.-E.; Validation, B.I.L., Y.B.-E., and D.I.; Formal Analysis, B.I.L. and D.I.; Investigation, B.I.L., Y.B.-E., and D.I.; Resources, Y.B.-E.; Writing-Original Draft Preparation, B.I.L.; Writing-Review and Editing, B.I.L., Y.B.-E., and D.I.; Visualization, D.I.; Supervision, Y.B.-E.; Project Administration, Y.B.-E.; Funding Acquisition, B.I.L. and Y.B.-E.

Funding: This research was funded by Holon Institute of Technology (HIT), Holon, Israel.

Conflicts of Interest: The authors declare no conflict of interest.

References

1. Zografopoulos, D.C.; Asquini, R.; Kriezis, E.E.; d'Alessandro, A.; Beccherelli, R. Guided-wave liquid crystal photonics. *Lab Chip* **2012**, *12*, 3598–3610. [CrossRef]
2. Khoo, I.C. *Liquid Crystals*, 2nd ed.; Wiley: Hoboken, NJ, USA, 2007.
3. Lazarev, G.; Chen, Po.; Strauss, J.; Fontaine, N.; Forbes, A. Beyond the display: Phase-only liquid crystal on Silicon devices and their applications in photonics. *Opt. Express* **2019**, *27*, 16206–16249. [CrossRef] [PubMed]
4. Frisken, S.; Clarke, I.; Poole, S. Technology and applications of liquid crystal on Silicon (LCoS) in telecommunications. In *Optical Fiber Telecommunications VIA*, 6th ed.; Kaminov, I.P., Li, T., Willner, A.E., Eds.; Academic Press—Elsevier: Oxford, UK, 2013; pp. 709–742.
5. Li, M.; Zhang, L.; Tong, L.-M.; Dai, D.-X. Hybrid silicon nonlinear photonics. *Photonics Res.* **2018**, *6*, B13–B22. [CrossRef]
6. Whinnery, J.R.; Hu, C.; Kwon, Y.S. Liquid-crystal waveguides for integrated optics. *IEEE J. Quantum Electron.* **1977**, *13*, 262–267. [CrossRef]
7. d'Alessandro, A.; Bellini, B.; Donisi, D.; Beccherelli, R.; Asquini, R. Nematic liquid crystal optical channel waveguides on Silicon. *IEEE J. Quantum Electron.* **2006**, *42*, 1084–1090. [CrossRef]
8. Donisi, D.; Bellini, B.; Beccherelli, R.; Asquini, R.; Gilardi, G.; Trotta, M.; d'Alessandro, A. A switchable liquid-crystal optical channel waveguide on Silicon. *IEEE J. Quantum Electron.* **2010**, *46*, 762–768. [CrossRef]
9. Cos, J.; Ferré-Borrull, J.; Marsal, L.F. Tunable waveguides based on liquid-infiltrated silicon photonic crystals. *Phys. Status Solidi C* **2011**, *8*, 1075–1078. [CrossRef]
10. Cai, D.-P.; Pan, H.-Y.; Tsai, J.-F.; Chiu, H.-K.; Nian, S.-C.; Chang, S.H.; Chen, C.-C.; Lee, C.-C. Liquid crystal infiltrated waveguide with distributed Bragg reflectors. *Opt. Mater. Express* **2011**, *1*, 1471–1477. [CrossRef]
11. Huang, C.-C. Solving the full anisotropic liquid crystal waveguides by using an iterative pseudospectral-based eigenvalue method. *Opt. Express* **2011**, *19*, 3363–3378. [CrossRef]
12. Shenoy, M.R.; Sharma, M.; Sinha, A. An electrically controlled nematic liquid core waveguide with a low switching threshold. *J. Light. Technol.* **2015**, *33*, 1948–1953. [CrossRef]
13. d'Alessandro, A.; Martini, L.; Gilardi, G.; Beccherelli, R.; Asquini, R. Polarization-independent nematic liquid crystal waveguides for optofluidic applications. *IEEE Photonic Technol. Lett.* **2015**, *27*, 1709–1712. [CrossRef]
14. Khoo, I.C. Nonlinear optics of liquid crystalline materials. *Phys. Rep.* **2009**, *471*, 221–267. [CrossRef]
15. de Gennes, P.G.; Prost, J. *The Physics of Liquid Crystals*, 2nd ed.; Oxford University Press: New York, NY, USA, 1993.
16. Stephen, M.J.; Straley, J.P. Physics of Liquid Crystals. *Rev. Mod. Phys.* **1974**, *46*, 617–704. [CrossRef]
17. de Gennes, P.G. *The Physics of Liquid Crystals*; Clarendon Press—Oxford: Wotton-under-Edge, UK, 1974.
18. Chandrasekhar, S. *Liquid Crystals*, 2nd ed.; Cambridge University Press: Cambridge, UK, 1992.
19. Liao, Y.; Clark, N.A.; Pershan, P.S. Brillouin scattering from smectic liquid crystals. *Phys. Rev. Lett.* **1973**, *30*, 639–641. [CrossRef]
20. Ricard, L.; Prost, J. "Second sound" propagation and the smectic response function. *J. Phys. Colloq.* **1979**, *40*, C3-83–C3-86. [CrossRef]
21. Ricard, L.; Prost, J. Critical behavior of second sound near the smectic A nematic phase transition. *J. Phys.* **1981**, *42*, 861–873. [CrossRef]
22. Qian, S.; Iannacchione, G.S.; Finotello, D. Critical behavior of a smectic-A to nematic phase transition imbedded in a random network of voids. *Phys. Rev. E* **1998**, *57*, 4305–4315. [CrossRef]
23. Khoo, I.C. Extreme nonlinear optics of nematic liquid crystals. *J. Opt. Soc. Am. B* **2011**, *28*, A45–A55. [CrossRef]
24. Giallorenzi, T.G.; Weiss, J.A.; Sheridan, J.P. Light scattering from smectic liquid-crystal waveguides. *J. Appl. Phys.* **1976**, *47*, 1820–1826. [CrossRef]
25. Ma, L.-L.; Hu, W.; Zheng, Z.-G.; Wu, S.-B.; Chen, P.; Li, Q.; Lu, Y.-Q. Light-activated liquid crystalline hierarchical architecture toward photonics. *Adv. Opt. Mater.* **2019**. [CrossRef]
26. Kventsel, G.F.; Lembrikov, B.I. Two-wave mixing on the cubic non-linearity in the smectic A liquid crystals. *Liq. Cryst.* **1994**, *16*, 159–172. [CrossRef]
27. Kventsel, G.F.; Lembrikov, B.I. Stimulated light scattering in smectic A liquid crystals. *Liq. Cryst.* **1995**, *19*, 21–37. [CrossRef]

28. Kventsel, G.F.; Lembrikov, B.I. Self-focusing and self-trapping in smectic A liquid crystals. *Mol. Cryst. Liq. Cryst.* **1995**, *262*, 629–643. [CrossRef]
29. Kventsel, G.F.; Lembrikov, B.I. The four-wave mixing and the hydrodynamic excitations in smectic A liquid crystals. *Mol. Cryst. Liq. Cryst.* **1995**, *262*, 591–627. [CrossRef]
30. Kventsel, G.F.; Lembrikov, B.I. Second sound and nonlinear optical phenomena in smectic A liquid crystals. *Mol. Cryst. Liq. Cryst.* **1996**, *282*, 145–189. [CrossRef]
31. Lembrikov, B.I. Light interaction with smectic A liquid crystals: Nonlinear effects. *HAIT J. Sci. Eng.* **2004**, *1*, 306–347.
32. Lembrikov, B.I.; Ben-Ezra, Y. Surface plasmon polariton (SPP) interactions at the interface of a metal and smectic liquid crystal. In Proceedings of the 17th International Conference on Transparent Optical Networks (ICTON 2015), Budapest, Hungary, 5–9 July 2015; We.C4.4, 1-4.
33. Lembrikov, B.I.; Ben-Ezra, Y.; Ianetz, D. Stimulated scattering of surface plasmon polaritons (SPPs) in smectic A liquid crystal. In Proceedings of the 18th International Conference on Transparent Optical Networks (ICTON-2016), Trento, Italy, 10–14 July 2016; We.B4.2, 1-4.
34. Lembrikov, B.I.; Ianetz, D.; Ben-Ezra, Y. Metal/Insulator/Metal (MIM) plasmonic waveguide containing a smectic A liquid crystal (SALC) layer. In Proceedings of the 19th International Conference on Transparent Optical Networks (ICTON 2017), Girona, Catalonia, Spain, 2–6 July 2017; Tu.A4.3, 1-4.
35. Lembrikov, B.I.; Ianetz, D.; Ben-Ezra, Y. Nonlinear optical phenomena in Silicon-Smectic A Liquid Crystal (SALC) waveguiding structures. In Proceedings of the 20th International Conference on Transparent Optical Networks (ICTON 2018), Bucharest, Romania, 1–5 July 2018; Mo.D4.1, 1-4.
36. Lembrikov, B.I.; Ianetz, D.; Ben-Ezra, Y. Nonlinear optical phenomena in smectic A liquid crystals. In *Liquid Crystals—Recent Advancements in Fundamental and Device Technologies*; Choudhury, P.K., Ed.; InTech: Rijeka, Croatia, 2018; pp. 131–157.
37. Shen, Y.R. *The Principles of Nonlinear Optics*; Wiley: Hoboken, NJ, USA, 2003.
38. Suhara, T.; Fujimura, M. *Waveguide Nonlinear-Optic Devices*; Springer: Berlin, Germany, 2003.
39. Haus, H.A. *Waves and Fields in Optoelectronics*; Prentice Hall: Upper Saddle River, NJ, USA, 1984.
40. Moloney, J.V.; Newell, A.C. *Nonlinear Optics*; Westview Press: Boulder, CO, USA, 2004.
41. Vagner, I.D.; Lembrikov, B.I.; Wyder, P. *Electrodynamics of Magnetoactive Media*; Springer: Heidelberg, Germany, 2004.

© 2019 by the authors. Licensee MDPI, Basel, Switzerland. This article is an open access article distributed under the terms and conditions of the Creative Commons Attribution (CC BY) license (http://creativecommons.org/licenses/by/4.0/).

Article

Ultrasonic Influence on Plasmonic Effects Exhibited by Photoactive Bimetallic Au-Pt Nanoparticles Suspended in Ethanol

Eric Abraham Hurtado-Aviles [1], Jesús Alejandro Torres [2], Martín Trejo-Valdez [3], Christopher René Torres-SanMiguel [1], Isaela Villalpando [4] and Carlos Torres-Torres [1,*]

1. Sección de Estudios de Posgrado e Investigación, Escuela Superior de Ingeniería Mecánica y Eléctrica Unidad Zacatenco, Instituto Politécnico Nacional, Ciudad de Mexico 07738, Mexico; ericabrahamh@gmail.com (E.A.H.-A.); ctorress@ipn.mx (C.R.T.-S.M.)
2. Academia de Ciencias (Acústica), Escuela de Laudería, Instituto Nacional de Bellas Artes y Literatura, Querétaro 76000, Mexico; jesusalejandrott@yahoo.com.mx
3. Escuela Superior de Ingeniería Química e Industrias Extractivas, Instituto Politécnico Nacional, Ciudad de Mexico 07738, Mexico; martin.trejo@laposte.net
4. Centro de Investigación para los Recursos Naturales, Salaices, Chihuahua 33941, Mexico; isaelav@hotmail.com
* Correspondence: crstorres@yahoo.com.mx or ctorrest@ipn.mx

Received: 6 April 2019; Accepted: 16 May 2019; Published: 3 June 2019

Abstract: The optical behavior exhibited by bimetallic nanoparticles was analyzed by the influence of ultrasonic and nonlinear optical waves in propagation through the samples contained in an ethanol suspension. The Au-Pt nanoparticles were prepared by a sol-gel method. Optical characterization recorded by UV-vis spectrophotometer shows two absorption peaks correlated to the synergistic effects of the bimetallic alloy. The structure and nanocrystalline nature of the samples were confirmed by Scanning Transmission Electron Microscopy with X-ray energy dispersive spectroscopy evaluations. The absorption of light associated with Surface Plasmon Resonance phenomena in the samples was modified by the dynamic influence of ultrasonic effects during the propagation of optical signals promoting nonlinear absorption and nonlinear refraction. The third-order nonlinear optical response of the nanoparticles dispersed in the ethanol-based fluid was explored by nanosecond pulses at 532 nm. The propagation of high-frequency sound waves through a nanofluid generates a destabilization in the distribution of the nanoparticles, avoiding possible agglomerations. Besides, the influence of mechanical perturbation, the container plays a major role in the resonance and attenuation effects. Ultrasound interactions together to nonlinear optical phenomena in nanofluids is a promising alternative field for a wide of applications for modulating quantum signals, sensors and acousto-optic devices.

Keywords: plasmonic nanoparticles; nonlinear acousto-optics; nanofluids; ultrasonic sensors

---

1. Introduction

The nonlinear optical (NLO) properties of nanomaterials are attractive, due to the possibility of their use in different technological fields [1]. In recent years, a great diversity of nanostructure thin films [2] has been reported observing different NLO properties which are attributable to the quantum confinement [3]. In particular, metal nanoparticles (NPs) have attracted great interest specially when are suspended in different based fluids [4], due to their physical and chemical characteristics sensitive to the local environment [5]. Also, nanofluids appear to be suitable for biomedical applications as molecular diagnostics, delivering sensing and bioimaging [6].

Diverse colloidal suspensions of metallic NPs have been the cornerstone of significant studies devoted to photonic [7], opto-electronic [1], electro-optical [8], and all-optical functions [9]. Multimetallic NPs display a wide range of interesting and potentially improved characteristics in comparison to monometallic NPs [10]. The superior features exhibited by multimetallic NPs are attributable to the Surface Plasmon Resonance (SPR) provided by their different components in a nanostructured configuration [11]. For example, bimetallic Gold and Platinum NPs (Au-Pt NPs) exhibit two peaks at the linear regime, which make them attractive for suitable catalyst [12] and electrochemical sensors. The formation of bimetallic alloys, such as core-shell and clusters, depends on the structure and size of the core [13], thus, they are widely influenced by the enhancement of the catalytic activity related to SPR effects [14]. Au-Pt NPs have been synthesized by different techniques as simple electrodeposition [15] and environmentally friendly procedures [16]; besides, it has been revealed the possibility to enhance their characteristics by tuning the molar ratios [17]. It must be noticed that the processing route for the preparation of the NPs plays critical roles in the size regime, morphology and distribution in comparison to the monometallic NPs [18]. Therefore, Au-Pt NPs emerge as highly promising materials for detection devices [19], and for biosensor signal processing owing to their improved physical, optical, and electromechanical properties [20].

The NLO response of thin films and NPs suspensions has been measured by different multiphoton absorption techniques [21]. Particularly, the NLO effects seem to be enhanced by tuning the optical field [22], which has been experimentally demonstrated in a wide range of nano-optic [23], acousto-optic [24], and acousto-plasmonic interactions.

Acoustic waves, such as infrasound, sound, and ultrasound (US) can be propagated as a mechanical disturbance in the matter, being potentially useful since they enable the ability to induce and control low-dimensional materials. Acoustic manipulation of metallic NPs is also allowed, due to the displacement of the media's molecules [25]. Acoustic vibrations, which have been used for enhanced-spectroscopy applications, currently have drawn interest from the NLO community [26]. Recent studies demonstrated the vibrations powered by sound in a liquid medium can generate autonomous motion, originated by nanorods trapped in an acoustic field [27]. Therefore, there is considerable interest from sound-light interaction, specially by mixing US and optical waves as a result of the rapid modulation and deflection of the light beams, as well as the demand for more general optical processing [28]. The interactions between US and nonlinear optical signals both open interesting opportunities for technological branches ranging from optics to photonics [29].

Also, multi-wave interactions promise potential for applications as acousto-optic modulation and compact photonic devices [30], as well as breast cancer diagnosis candidates for acousto-optic imaging in biomedicine [31]. This is because, while techniques that deal with the problem of image generation from diffuse photons suffer low signal-to-noise ratio, the light which has passed through a scattering medium can be easily detected and localized. Therefore, even a three dimensional localization of breast tumors and their characterizations can be obtained through ultrasonically-controlled signals.

Particular properties derived from quantum confinement and the interaction between nanostructures has been indicated in metal nanostructures [32,33]. Some interesting aspects of plasmonic materials are their large absorption cross-sections and strongly localized electromagnetic fields [34]. Remarkably, the linear and NLO effects play an important role in the vibrational mode driven by acousto-plasmonic coherent control and result in a change of the refractive index [35]. In light-sound interactions, the modification of the optical refractive index is due to the particle's compression and additional refraction effects can be originated by the sound wave propagation through a fluid media [36]. It has been demonstrated that plasmonic metamaterial-based sensors for US detection, present advantages derived from their strong sensitivity to the refractive index of their surroundings [37]. Besides, opto-mechanical nonlinearities exhibited by plasmonic metamaterials can be modulated by third-order NLO parameters [38]. Plasmonic nanostructures composed of metallic and dielectric media, exhibit tunable plasmon properties associated with the collective oscillations of free electrons [39]. And another important fact is the possibility to create hybrid systems by the combination of independent elements with the aim to enhance their characteristics and applications [40]. In view of all these considerations, within this work have been analyzed optical effects that can be related to US and plasmonic phenomena. We presented variable contributions from US to NLO transmittance in single-beam and TWM experiments; these results are a consequence of the synergistic potential of plasmonics and US for modulation of optical absorption and refractive index; respectively. Specifically, in this work, an attempt has been made for further investigate the transmittance of NLO signals may receive an opposite influence from multiphotonic absorption and the optical Kerr effect when the propagation of US takes place in a nonlinear sample. Some examples that illustrate the importance of modulating optical signals by ultrasonic frequencies are related to quantum plasmonic sensors assisted by NLO effects [41] and microfluids with properties governed by the incorporation of plasmonic nanostructures and acoustic signals [42]. In this direction, the main purpose of this work is to evaluate the influence of ultrasonic signals in the plasmonic response exhibited by nanocolloidal solutions. Herein is reported a study about the ultrasonically activated modification on the plasmonic and absorption effects exhibited by bimetallic gold-platinum nanoparticles. Different contributions related to mechanical phenomena and NLO processes were observed. The significance of this study comes from the clear indication of enhancement and inhibition of the optical transmittance related to plasmonic sensing. This work highlights the attractive nonlinear optical characteristics associated with the samples for developing photoactive advanced materials.

2. Materials and Methods

2.1. Sample Synthesis

Colloidal bimetallic Au-Pt NPs were synthesized through a sol-gel technique accordingly to a previous report [43]. Briefly, the process was carried out as follows; the sol-gel TiO_2 was obtained from Titanium i-propoxyde [$Ti(OC_3H_7)_4$] used as a precursor with a concentration C = 0.05 mol/L, pH = 1.25, was dissolved in a water/alkoxide solution with a molar ratio 0.8. Then, standard solutions of Au and Pt precursors with an equivalent nominal metal concentration of 1000 mg/L each were used. The mixture of (Au + Pt)/$Ti(OC_3H_7)_4$ revealed a molar ratio of 0.76% (mol/mol) in a total volume of 11.5 mL. The resulting solution was kept in darkness for a week before nucleation with the aim to extend their photo-response, improve their catalytic activity, stabilize the sample, and use it several times [44]. Finally, an ultraviolet (UV) light reactor was used in the photocatalytic processes for the preparation of the bimetallic sample.

2.2. Morphology Characterization

The morphology and size distribution of the sample were both studied by Transmission Electron Microscopy and High-Resolution Transmission Electron Microscopy (TEM and HRTEM, FEI Titan 80-300, JEM-ARM200CF) techniques. For the microscopic measurements, a drop of the studied suspension was placed on a carbon-coated Cu grid, it was allowed to dry overnight at room temperature prior to the optical explorations. TEM micrographs were acquired using a probe aberration-corrected microscope, recorded by a Gatan Ultrascan 1000 xP digital camera and operated at an acceleration voltage of 80–200 kV. Besides, Energy-dispersive X-ray analysis (EDX; JECL JSM-7800F) were undertaken to clarify the chemical composition of the Au-Pt NPs.

2.3. Nanosecond Optical Single-Beam Transmittance

The bimetallic Au-Pt NPs were explored by an Nd:YAG nanosecond laser source (Continuum Model SL II-10) with 4 ns pulse duration at 532 nm wavelength, repetition rate of 10 Hz, and linear polarization. Figure 1 illustrates the scheme of a single-beam experiment setup performed in our laboratory with the aim to investigate the nonlinear optical response of the solution.

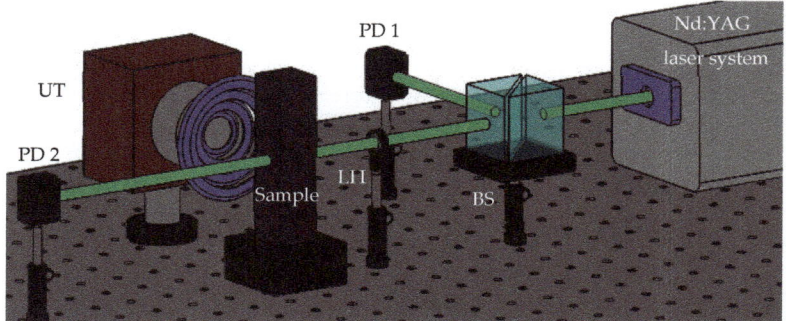

Figure 1. Experimental setup for single-beam transmittance measurements influenced by US.

2.4. Acousto-Optic Explorations

In order to study the US influence on the Au-Pt NPs, the acousto-optical exploration was also investigated with the arrangement schematized in Figure 1. The excitation beam was divided by means of a beam splitter, BS, with the purpose of monitoring power emission of the laser system by a photodiode detector (PD-1) (Newport, Model: 818-BB-21). In the experiment, both light and US signals were simultaneously propagated on the colloidal sample. The optical irradiation was focused to a small spot by using a quartz lens (LH) (Newport, Plano-Convex Lens) and increased gradually while the acoustic square wave remained continuous.

A home-built electronic circuit model by using ultrasonic transducer (UT) was employed to generate US. In accordance with the specification supplied by the manufacturer, the sensor operates in a frequency range of 40.0 ± 1.0 kHz. In order to start the operation of the UT, it must be selected a high pulse of high voltage (5V) during at least 10 µs, then, the trigger pin will transmit out 8 cycles of ultrasonic burst at 40 kHz and an ultrasonic burst is going to be reflected. The echoes from the target cannot be ignored, since they are essential to the circuit model to consider the following burst of the 40 kHz square-wave emission.

The energy of the transmitted optical beam was measured by a digital oscilloscope (Teledyne LeCroy, WaveSurfer 3054) using a photodiode detector (PD-2) (Newport, Model: 818-BB-21). Finally, both optical and ultrasonic signals were compared to evaluate the influence of the ultrasonic disturbance.

2.5. Two-Wave Mixing Experiment Influenced by US

A two-wave mixing (TWM) experiment schematized in Figure 2 was carried out to identify the vectorial nonlinear optical response of the sample. In this technique, the laser source linearly polarized with a Gaussian profile was focused by using a 75 cm focal-length lens (LH). The pulse energy of 115 mJ provided by the Nd:YAG laser system was divided into two beams by a cubic beam splitter (BS) both featuring the same light polarization direction. The two beams, pump and probe, interfere into the optical cell with an optical irradiance relation of 10:1, each beam path was guided to the sample by high-energy laser mirrors (Newport, Dual-wavelength mirror for 1064 and 532 nm. 2.0 in. Diameter), (M 1-2). An achromatic half-wave plate (WP), (Newport, Quartz-MgF2), was inserted in the pump arm to change the direction of the linear polarization with a Newport, Calcite Polarizer (P). The transmittance of the sample for both beams was measured along the laser propagation direction by PIN photodetectors (PD 1-2) (Newport, Model: 818-BB-21). The temporal and spatial superposition of the beams was verified by the optimization of the position of the mirrors considering the Kerr transmittance of the probe beam with the influence of the pump beam interacting with the sample.

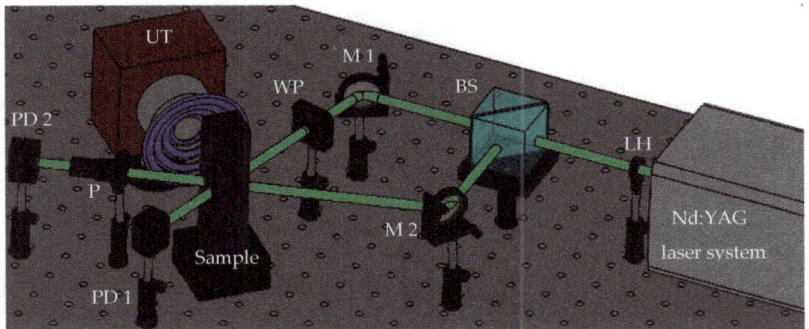

Figure 2. Acousto-optic explorations by the assistance of the two-wave mixing configuration.

For further investigate the coupling between the pump and the probe beams, we considered the right and left circular components of the electric field as $E_\pm$ and $E_\mp$. Then, we applied the finite-differences method to solve the Maxwell's wave equation given by [45]:

$$\nabla^2 E_\pm = -\frac{n_\pm^2 \omega^2}{c^2} E_\pm, \tag{1}$$

where c is the velocity of light in the vacuum and the optical frequency is defined by ω. We assume, that both waves propagate the sample in the same plane. In our representation, is convenient to express the refractive index taking into account the approximation [45]:

$$n_\pm^2 = n_0^2 + 4\pi\left(A|E_\pm|^2 + (A+B)|E_\mp|^2\right), \tag{2}$$

here n_0 is the weak-field refractive index. $A = \chi^{(3)}_{1122}$ and $B = \chi^{(3)}_{1212}$ are the independent components of the third-order susceptibility tensor $\chi^{(3)}$.

2.6. Modulation of Plasmonic Signals by Ultrasonic Effects

A path length of 1 cm in a quartz cuvette was heuristically filled with the dispersion of Au-Pt NPs to explore the linear and nonlinear optical response of the samples. UV-Visible optical absorbance spectra were acquired by using an Ocean Optics UV-vis fiber optic-based spectrometer (USB 2000+XR1-ES), equipped with deuterium-halogen light source. Additionally, in order to modulate the influence of the optical response, an Ultrasonic Transducer (UT-1 and UT-2) situated above the optical quartz cuvette was used for delivering US waves in the sample. The prepared solution was scrutinized in the wavelength range from 200 nm to 900 nm.

The modulation of the plasmonic signals was achieved by ultrasonic interactions; the experimental setup is shown in Figure 3. Near-resonance excitations were promoted by monochromatic light of $\lambda = 532$ nm provided by a horizontally polarized Nd:YVO$_4$ pulsed laser (Spectra-Physics Explorer® One™ XP). The laser with constant energy of 7.3 µJ was split into two beams of equal energy by means of a BS; one beam was focused on the sample using a Newport, Plano-Convex Lens with a focal length of 20 cm, while the other arm beam was used as a reference.

The sample was disturbed with US combined with the stable pulse energy of the laser source. In this case, the ultrasonic frequency was modulated by using a pair of different transducers (UT 1-2); their operation was previously described (Section 2.4). The optical quartz cuvette with the colloidal solution of Au-Pt NPs were placed between the two UT 1-2 separated at the same distance.

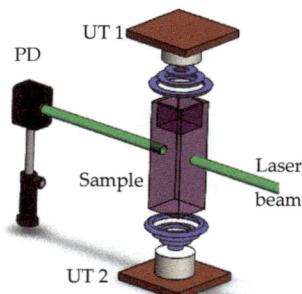

Figure 3. Single-beam transmittance set-up based in an Nd:YVO$_4$ pulsed laser.

However, in order to obtain the phase difference frequency, we used different 40 kHz square-wave signal-each burst in 20 periods. The output light beam profile of the transmitted light was collected by a photodiode (PD) (Newport, Model: 818-BB-21) and processed by the digital oscilloscope.

In this work, the modulation of optical signals was obtained by the combination of light with US. By using an UV-vis spectroscope is proposed to modulate the magnitude of the peak in the absorption band related to the plasmonic response of the sample interacting with ultrasonic signals. Furthermore, when the transmittance of a sample changes with the incident irradiance, then nonlinear optical effects are involved. Third-order NLO phenomena are strongly dependent on the plasmonic response of a sample; while nonlinear refraction and nonlinear absorption can be responsible for third-order nonlinear optical processes. By conducting single-beam transmittance experiments, we proposed that multi-photonic absorption processes together to ultrasonic signals is possible to increase the modulation in the transmittance of a plasmonic sample. On the contrary, due to the TWM experiments, where an induced nonlinear refractive index can be demonstrated, it is indicated the possibility to decrease the transmittance of the sample by the influence of US signals responsible for a modification in the refractive index of the sample.

3. Results

3.1. Characterization of the Synthesized Au-Pt NPs: TEM and EDX

The representative TEM image, shown in Figure 4, confirms the size and distribution of the sample. The diameter of the distributed Au-Pt NPs ranged between 9–13 nm, exhibiting an average standard derivation close to 10.5 nm. The statistics were carried out by measuring different individual particle diameters from randomly acquired TEM images. TEM can be used to perform the chemical mapping to show the Pt and Au layer on the nanoparticles. However, we guarantee the bimetallic nature in the composition of the sample by statistical EDX analysis.

The formation of Au-Pt NPs was confirmed by EDX studies mapped on Figure 5. The chemical composition of the alloy structure shows that Au is the predominant element, although the same amounts of standard solutions of Au and Pt precursors were used. The stronger photocatalytic activity exhibited by Pt seems to be responsible for this result. The atomic relation between Au and Pt in the NPs was about 10:1; respectively. EDX pattern also reveals the presence of other peaks associated from the copper substrate. We identified that increasing the Au content in Au-Pt alloys originates larger NPs which can be useful in different fields [46].

Figure 4. High-Resolution TEM image of the Au-Pt NPs embedded in TiO_2 film studied.

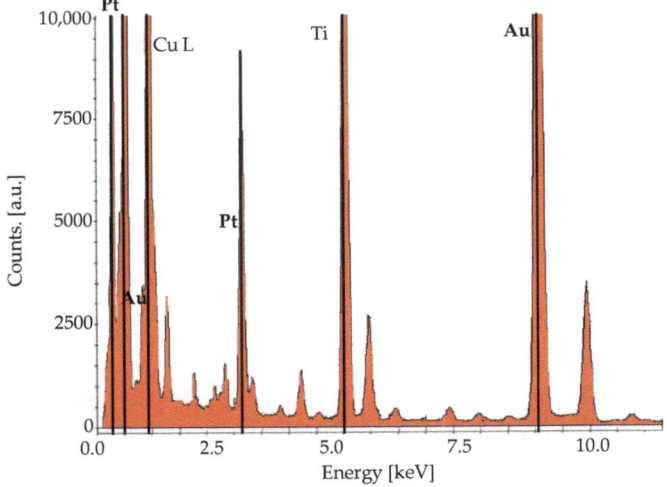

Figure 5. Chemical composition of the synthesized bimetallic sample studied by EDX.

3.2. Single-Beam Transmittance Measurements Affected by Ultrasonic Vibrations

Acousto-optical explorations were undertaken considering that both light and acoustic waves were propagated together through the colloidal Au-Pt NPs. Figure 6 summarizes the result of the experimental setup using the Nd:YAG laser system. Initially, we measured the optical single-beam transmittance exhibited by the sample. Then, the simultaneous interaction of both optical and ultrasonic signals was considered; the blue circles' markers depicted the final contribution of the US in the optical transmittance.

It should be noted that the interaction of both optical and ultrasonic waves influences the stability of the bimetallic colloidal solution with Au-Pt NPs. The modification of the optical transmittance as a function on US signal can be explained by considering that the US frequency is capable to modify the optical and the refractive index; since both are strongly dependent of the vibrational mode of the sample. One of the effects that would especially benefit from the contribution of the US, is that the sedimentation process decreases, due to the high-mechanical oscillations. Besides, it could prevent the appearance of agglomerations and accumulated NPs exposed to electrostatic effects. Herein, the presence of bimetallic NPs within a host liquid will not only lead to the modification of SPR effects that may improve linear and nonlinear absorption, but it could modulate their response owing to the US frequency, yielding to a larger optical absorption enhancement [47].

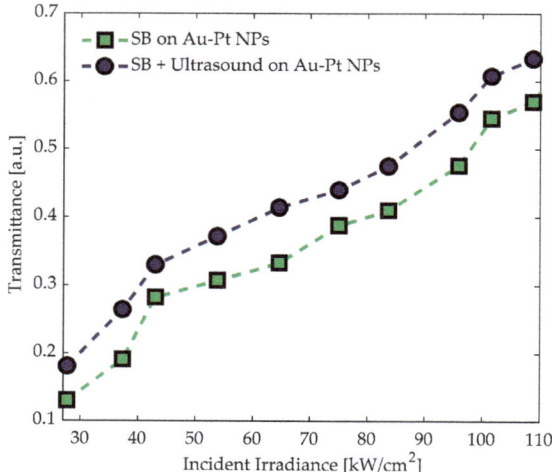

Figure 6. Change in the single-beam transmittance by the ultrasonic waves propagated through the Au-Pt NPs suspended in ethanol.

The morphology and particle size distribution are not affected by the precursor; however, a shift of the optical absorption spectra could be presented by decreasing the grain size [48]. Besides, solvent-metal interactions are important in NLO effects, due to the refractive index plays a prominent role in third-order optical nonlinearities [49].

In addition, the nonlinear optical properties depend on the absorption cross-section and transition energy level, as a result, the nonlinear optical transmittance increases or decreases as a function on light irradiance. In our case, the monotonic increase in transmittance, shown in Figure 6 can be considered as a signature of a typical saturable absorption dependent on incident irradiance.

3.3. TWM Measurements Affected by Ultrasonic Vibrations

The nonlinear optical response of the bimetallic Au-Pt NPs suspended in an optical cell was evaluated by a nanosecond TWM technique. In this case, the angle of polarization was controlled by means of a half-wave plate (WP), (Newport, Quartz-MgF2) placed in the pump arm before to irradiate the sample. The output of probe beam provided by the Nd:YAG laser system was measured using a PIN diode PD 2 (Newport, Model: 818-BB-21). The measurements were performed for two different cases, shown in Figure 7; the blue circles represent the typical angular-dependence of the interferometric optical measurements while the plotted square red markers correspond to the values with the presence of the ultrasonic waves. The magnitude in the optical transmittance of the sample was tuned by the superposition of the waves in order to obtain an interference fringe pattern in the nonlinear medium. The ultrasonic transducer was emitting short US pulses featuring 40.0 ± 1.0 kHz square waves at a level of about 120 dB. The nanoparticle density is strongly dependent on the US frequency [50]. Therefore, the refractive index and transmittance appear to be greatly influenced by the US action.

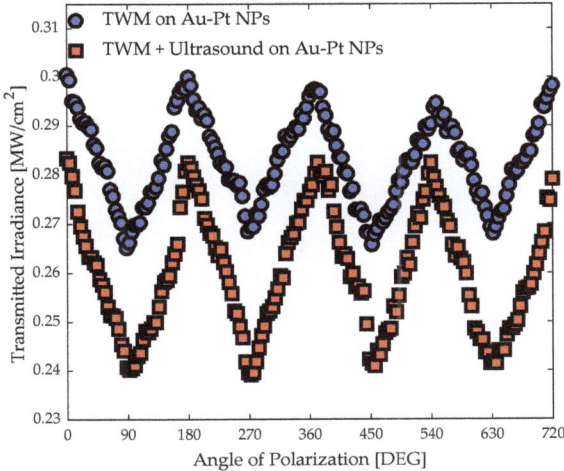

Figure 7. Transmitted irradiance as a function on the angle between the planes of polarization of the incident optical beams with and without an acousto-optic influence.

From the results illustrated in Figure 7, the transmitted irradiance pointed out a change in the optical nonlinearity of the sample derived by the influence of ultrasonic waves. In this case, we consider that the dynamic motion of the sample can generate an inhibition in the photoactive electronic response of the sample exposed to a near-resonance optical Kerr effect (OKE) excitation. The OKE technique has been reported to investigate the orientation dynamics of molecular liquids approaching a phase transition [51] and it has a strong dependence on the SPR properties exhibited by plasmonic NPs. The OKE manifests itself in a TWM interaction considering that a dependence on the optical irradiance can be also related to the refractive index of the media. In this TWM inside of a Kerr medium, a dependence on the host media could be remarkable because a strong influence of the US could originate changes in the density and in the induced birefringence. It is worth noting that the results plotted in Figure 7 are modulated by the angle of polarization, since the interference pattern associated with the two incident waves is dependent on the polarization of the waves. Therefore, it can be assumed that the plasmonic response of the samples is the main responsible for the modulation of the nonlinear optical transmittance controlled by the US frequency. Furthermore, it can be considered that changes in absorptive nonlinearities rising from the stationary interference pattern of the incident optical waves in the samples may contribute also to the change in the nonlinear refractive index [52]. The simultaneous participation of optical and ultrasonic interaction could be responsible for the modification of quantum confinement phenomena [11] which is related to the volume concentration, the size, and composition of the sample [18].

3.4. Ultrasonically-Controlled Plasmonic Signals by Bimetallic Au-Pt Nanoparticles Suspended in Ethanol

In Figure 8 are presented the UV-Visible absorption spectra of the sample with the participation of an ultrasonic wave excitation. These measurements represent experimental data acquired 10 different times in different regions of similar samples systematically studied. Gyroscopic behavior of the NPs seems to modulate almost uniformly the UV-vis spectra of the sample as it has been previously demonstrated [53]. In this case, the change in the linear absorption is due to a modification in the plasmonic response of the bimetallic sample. US waves are able to modulate the density in nanofluids and then, it can be expected an important modification in the collective electronic response of plasmonic samples. From the plot can be clearly seen that the colloids show a shift in the absorbance response attributed to the US wave propagation. The absorption peak presented in the UV region close to

300 nm corresponds to a plasmonic absorption of the Pt, while the absorption peak in the visible region close to 650 nm is related to the response of the Au integrating the NPs. The significance of these absorption bands is related to the selective absorptive behavior of the characteristic effects of the size, shape and distribution of each metal integrating the nanostructured sample.

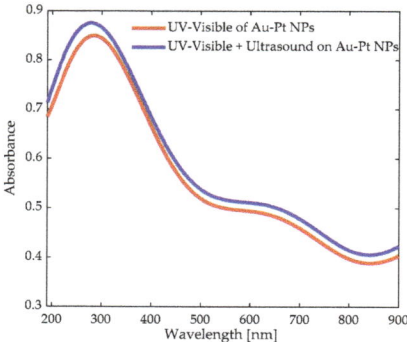

Figure 8. Representative UV-Visible absorption spectra of the studied samples showing a change derived by ultrasonic signals.

The UV-vis spectrum is associated with a random distribution of the dispersed NPs, and not only to those shown in the TEM image. Different bimetallic Au-Pt NPs alloys have been studied showing a change in the SPR peak by increasing the Au content, and due to the Au element is more electronegative than Pt [46], which originates a modification of its absorption properties and linear and nonlinear optical refraction. Hence, the possibility to change the concentration of the Au-Pt NPs exhibit unique extraordinary features induced by the plasmonic response can be attributable to the nanomechanic and nanophotonic properties of the sample.

Nanoscale effects are very important for instrumentation applications, such as plasmonic and nanophotonic sensors, regarding the high sensitivity associated with low-dimensional interactions. Also, it has been found that nanoscale phenomena are strongly dependent on the quantum confinement which appears to be improved by the participation of US frequency. The power and frequency of the US seems to be responsible for the modification of the plasmonic effects related to nanosystems [54].

Regarding the possibility to change the optical absorption of the NPs by the influence of an ultrasonic effect, the modification of the amplitude of near-resonance optical signals in the sample was considered. In Figure 9, the sawtooth waveform corresponds to the excitation of plasmonic signals induced by a 532 nm wavelength exciting the SPR associated with the Au atoms of the samples. The value of the single-beam transmittance measurement with and without the participation of an ultrasonic control was analyzed. In addition to this experiment, an ultrasonic signal was propagated from the transversal direction to the laser beam, which passed through the sample cell containing Au-Pt NPs in ethanol. The data plotted in red color were obtained by modulating the amplitude in the US frequency of 40 kHz. The changes in the y-axis of Figure 9 correspond to the transmittance when an Nd:YVO4 pulsed laser system irradiates the sample. The resulting data were obtained for two different cases; initially, a PIN diode (PD 2) at the output of measured the value when the laser irradiates the sample with a repetition period of T = 1/50 kHz = 0.02 ms. Then, the optical transmittance was measured in simultaneous propagation of the US square wave, delivered at a pulse repetition frequency of 40 kHz over a total exposure time of 5 s.

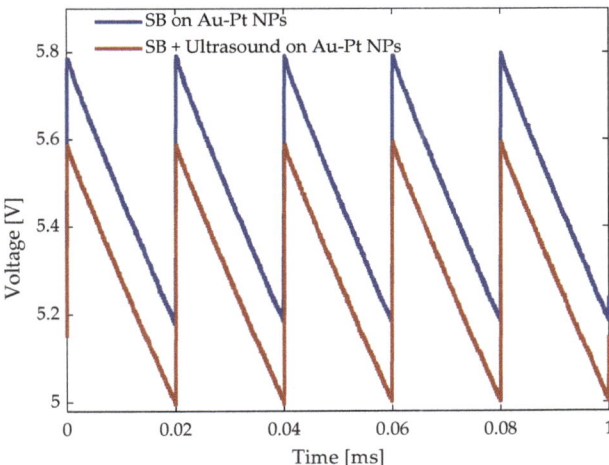

Figure 9. Amplitude vs time spectrum of single-beam transmittance measured at the output light with a pulse energy of 5.5 µJ.

One interesting observation is the monotonic decrease in the plasmonic and optical transmittance in a linear relationship with the US increase in good agreement with previous studies [43]. Low-level ultrasonic signals were systematically employed with repetitive results within an error bar below 1%. The possibility to change the bimetallic structure exhibited by the NPs under the influence of US waves was considered to be part of further investigations. Change in metallic structures has been previously reported [55] and also the effects of grain size decrease by US waves [56].

All the measurements were carried out in the presence of the US perturbation, also was performed a base data acquirement to observe the influence of the US in the studied Au-Pt NPs. We observed the highest contribution of the US in the optical response of the NPs when the US source was in the neighborhood of the surface of the sample; distances of several centimeters practically inhibit the influence of the US frequency in the optical effects evaluated. The modification of the optical transmittance as a function on US signal can be explained by considering that the US frequency is capable to modify the optical and the refractive index; since both are strongly dependent of the vibrational mode of the sample. The study of NPs and nanofluids are of great interest in nanofluidic electronic circuits to manipulate ions and create biomolecular diagnosis devices [1]. The idea of constructing some new devices based on the combination or rearrangement of diodes is found to be applicable in nanofluidics. p–n junction diodes are the elementary building blocks of most semiconductor devices [57]. In the feedback of all-nanoparticle logic systems, can be oppositely charged by using laminated nanoparticle layers. Nanostructured devices based on the advantages of plasmonic effects can be found in bipolar junction transistors, field-effect transistors or thyristors [58]. Doped p–n junctions have been created considering electrostatics instead of adding different chemical elements in single configurations [59]. A chemoelectronic circuit composed of Au NPs coated with different types of organic molecules has been recently suggested for the development of diodes/transistors. In this respect, the major interest is manipulating the ions' movement of the packed Au NPs in an inherent electric field [56].

Particularly, hybrid nanoscale acousto-optic systems have been proposed as acousto-optic sensors, acousto-optic wireless communication modulators and transducers based on graphene [60]. As well, Au NPs have been studied by using acousto-plasmonic interactions [26]. However, the optical and catalytic properties can be improved by combining two different metals at nanometric scale [61], in which, metal composition and shape provide unique functionality [62]. Among the different bimetallic NPs, Au-Pt NPs show higher electrocatalytic activity that represents widely promising functions in catalysis applications, such as a suitable catalyst for methanol oxidation and oxygen reduction reaction [63]. The enhanced photocatalytic activity is attributable to the synergy of the metal composition and the SPR effects.

It has been previously reported that third-order optical properties may conduct to a larger sensitivity to local changes in the refractive index compared to the commonly used linear localized SPR sensing [64]. It has been demonstrated that the optical transmission of nanoplasmonic materials becomes spectrally tunable by acoustic waves [65]. Moreover, monotonic changes in the sensitivity exhibited by nonlinear optical sensors have been indicated [43]. Contrastingly, in this work, we pointed out the possibility to increase or decrease the modulation of optical signals by linear or nonlinear optical effects; respectively, in plasmonic NPs.

The third-order nonlinear optical susceptibilities can be responsible for the change in the refractive index and the nonlinear absorption of colloidal solutions irradiated by light [66]. These changes in refractive index automatically modify the mechanical properties of the sample that depend on mass density. Specifically, light-sound interactions have been used for developing nonlinear acousto-optical modulation which depends on the resonance and the vibrational mode of the metal nanostructures [30]. We noted that the influence of US in nonlinear optical properties can generate a modification in optical transmittance that is opposite to the modification of the propagation of light in the linear regime. Our results are in good agreement with nonlinear optical effects in liquids that demonstrate the possibility to modify the refractive index [67] taking into account that plasmonic resonances show considerable changes in optical properties exhibited by nanocomposites [68]. Comparatively, herein is reported the observation of ultrasonic influence and their potential to separately modulate with opposite behavior the linear and nonlinear optical effects in plasmonic NPs. In this work has been highlighted the important influence of simultaneous interaction of both optical and ultrasonic signals to control plasmonic phenomena.

4. Conclusions

A strong influence in the modulation of the optical and plasmonic properties exhibited by bimetallic NPs was demonstrated by using ultrasonic waves. Spectral and TWM experiments in an Au-Pt based nanofluid under the influence of ultrasonic signals were reported. A reduction in the magnitude of the nonlinear optical transmittance dependent on external mechanical perturbations was observed. Herein is identified that the US wave propagation in nanostructured materials can be useful for controlling nonlinear and plasmonic signals. From the point of view of quantum optics, NLO receives an important influence from plasmonics and mechanical effects. Then the study of the modification of nonlinear optical phenomena assisted by US signals promises to be effective in a wide range of applications for instrumentation and low-dimensional signal processing where mechanical and plasmonic actions occur. The ultrasonic and nonlinear optical behavior of bimetallic Au-Pt NPs can be considered as potential candidates to develop quantum operations, sensors and instrumentation devices in biomedicine.

Author Contributions: E.A.H.-A. carried out optical and ultrasonic experiments; J.A.T. contributed in the analysis related to acoustical effects, M.T.-V. is responsible for the preparation and structure characterization of the samples; C.R.T.-S.M. participated in discussions about the mechanical observations; I.V. analyzed the bimetallic nature of the samples, C.T.-T. elucidated the nonlinear properties and designed the paper. The manuscript was written through the contribution of all authors.

Funding: This research was funded by Consejo Nacional de Ciencia y Tecnología (CONACyT) (grant No. CB-2015-251201), Comisión de Operación y Fomento de Actividades Académicas, Instituto Politécnico (COFAA) (2019) and Instituto Politécnico Nacional (IPN) (SIP-2019).

Acknowledgments: The authors kindly acknowledge the financial support from the Instituto Politécnico Nacional, COFAA-IPN, and Consejo Nacional de Ciencia y Tecnología. The authors are also thankful to the Central Microscopy facilities of the Centro de Nanociencias y Micro y Nanotecnología del Instituto Politécnico Nacional.

Conflicts of Interest: The authors confirm that there are no conflicts of interest related to this research.

References

1. Huang, D.; Zheng, C.; Huang, L.; Xiukai, W.; Ling, C. Linear and nonlinear optical properties of ultrafine WO_3 nanorods. *Optik* **2018**, *156*, 994–998. [CrossRef]
2. Etminan, M.; Hajiesmaeilbaigi, F.; Koohian, A.; Motamedi, A.; Golian, Y. Nonlinear optical properties of pulsed laser deposited Cu and Zn single and double layer nanostructure thin films. *Thin Solid Films* **2016**, *615*, 1–7. [CrossRef]
3. Dai, H.; Zhang, L.; Wang, Z.; Wang, X.; Zhang, J.; Gong, H.; Han, J.-B.; Han, Y. Linear and nonlinear optical properties of silver-coated gold nanorods. *J. Phys. Chem. C* **2017**, *121*, 12358–12364. [CrossRef]
4. Türk, M.; Erkey, C. Synthesis of supported nanoparticles in supercritical fluids by supercritical fluid reactive deposition: Current state, further perspectives and needs. *J. Supercrit. Fluids* **2018**, *134*, 176–183. [CrossRef]
5. Pfeiffer, C.; Rehbock, C.; Hühn, D.; Carrillo-Carrion, C.; de Aberasturi, D.J.; Merk, V.; Barcikowski, S.; Parak, W.J. Interaction of colloidal nanoparticles with their local environment: The (ionic) nanoenvironment around nanoparticles is different from bulk and determines the physico-chemical properties of the nanoparticles. *J. R. Soc. Interface* **2014**, *11*, 20130931. [CrossRef]
6. Cordeiro, M.; Ferreira, C.F.; Pedrosa, P.; Lopez, A.; Baptista, P.V. Gold Nanoparticles for Diagnostics: Advances towards Points of Care. *Diagnostics* **2016**, *6*, 43. [CrossRef]
7. Karimzadeh, R.; Mansour, N. The effect of concentration on the thermo-optical properties of colloidal silver nanoparticles. *Opt. Laser Technol.* **2010**, *42*, 783–789. [CrossRef]
8. Han, J.; Freyman, M.C.; Feigenbaum, E.; Han, T.Y.J. Electro-optical device with tunable transparency using colloidal core/shell nanoparticles. *ACS Photonics* **2018**, *5*, 1343–1350. [CrossRef]
9. Gargiulo, J.; Violi, I.L.; Cerrota, S.; Chvátal, L.; Cortés, E.; Perassi, E.M.; Diaz, F.; Zemánek, P.; Stefani, F.D. Accuracy and Mechanistic Details of Optical Printing of Single Au and Ag Nanoparticles. *ACS Nano* **2017**, *11*, 9678–9688. [CrossRef]
10. He, W.; Wu, X.; Liu, J.; Hu, X.; Zhang, K.; Hou, S.; Zhou, W.; Xie, S. Design of AgM Bimetallic Alloy Nanostructures (M = Au, Pd, Pt) with Tunable Morphology and Peroxidase-Like Activity. *Chem. Mater.* **2010**, *22*, 2988–2994. [CrossRef]
11. Li, C.-H.; Li, M.-C.; Liu, S.-P.; Jamison, A.C.; Lee, D.; Lee, T.R.; Lee, T.-C. Plasmonically Enhanced Photocatalytic Hydrogen Production from Water: The Critical Role of Tunable Surface Plasmon Resonance from Gold-Silver Nanoshells. *ACS Appl. Mater. Interfaces* **2016**, *8*, 9152–9161. [CrossRef]
12. Piña-Díaz, A.J.; Trejo-Valdez, M.; Morales-Bonilla, S.; Torres-San Miguel, C.R.; Martínez-González, C.L.; Torres-Torres, C. Nonlinear mechano-optical transmittance controlled by a rotating TiO2 thin solid film with embedded bimetallic Au-Pt nanoparticles. *J. Nanomat.* **2017**, *2017*, 2918509. [CrossRef]
13. Mahmood, A.; Shah, A.; Shahzad, S.; Aftab, S.; Latif-ur-Rahman, N.J.; Shaha, A.H. Monitoring of an Anti-Ulcer Drug Rabeprazole Using Au-Pt Bimetallic Alloy Nanoscale Electrochemical Sensor. *J. Electrochem. Soc.* **2017**, *164*, H413–H419. [CrossRef]
14. Haldar, K.K.; Kundu, S.; Patra, A. Core-Size-Dependent Catalytic Properties of Bimetallic Au/Ag Core–Shell Nanoparticles. *ACS Appl. Mater. Interfaces* **2014**, *6*, 21946–21953. [CrossRef]
15. Zhao, L.Y.; Thomas, J.F.; Heinig, N.F.; Abd-Ellah, M.; Wang, X.Y.; Leung, K.T. Au–Pt alloy nanocatalysts for electro-oxidation of methanol and their application for fast-response non-enzymatic alcohol sensing. *J. Mater. Chem. C* **2014**, *2*, 2707–2714. [CrossRef]
16. Ye, W.; Yu, J.; Zhou, Y.; Gao, D.; Wang, D.; Wang, C.; Xue, D. Green synthesis of Pt–Au dendrimer-like nanoparticles supported on polydopamine-functionalized graphene and their high performance toward 4-nitrophenol reduction. *Appl. Catal. B Environ.* **2016**, *181*, 371–378. [CrossRef]
17. Mistry, H.; Reske, R.; Strasser, P.; Cuenya, B.R. Size-dependent reactivity of gold-copper bimetallic nanoparticles during CO_2 electroreduction. *Catal. Today* **2017**, *288*, 30–36. [CrossRef]
18. Wang, S.; Kristian, N.; Jiang, S.; Wang, X. Controlled synthesis of dendritic Au@Pt core-shell nanomaterials for use as an effective fuel cell electrocatalyst. *Nanotechnology* **2009**, *20*, 025605. [CrossRef] [PubMed]
19. Wang, W.; Zou, Y.; Yan, J.; Liu, J.; Chen, H.; Li, S.; Zhang, L. Ultrasensitive colorimetric immunoassay for hCG detection based on dual catalysis of Au@Pt core–shell nanoparticle functionalized by horseradish peroxidase. *Spectrochim. Acta Part A* **2018**, *193*, 102–108. [CrossRef]

20. Hurtado-Aviles, E.A.; Torres, J.A.; Trejo-Valdez, M.; Romero-Ángeles, B.; Villalpando, I.; Torres-Torres, C. Amplitude-modulated acoustic waves by nonlinear optical signals in bimetallic Au-Pt nanoparticles and ethanol based nanofluids. *J. Mol. Liq.* **2018**, *263*, 288–293. [CrossRef]
21. Carriles, R.; Schafer, D.N.; Sheetz, K.E.; Field, J.J.; Cisek, R.; Barzda, V.; Sylvester, A.W.; Squier, J.A. Invited review article: Imaging techniques for harmonic and multiphoton absorption fluorescence microscopy. *Rev. Sci. Instrum.* **2009**, *80*, 081101. [CrossRef] [PubMed]
22. Bijeesh, M.M.; Shakhi, P.K.; Varier, G.K.; Nandakumar, P. Tuning the nonlinear optical absorption in Au/BaTiO3 nanocomposites with gold nanoparticle concentration. *Opt. Laser Technol.* **2018**, *102*, 207–212. [CrossRef]
23. Torres-Torres, C.; Tamayo-Rivera, L.; Rangel-Rojo, R.; Torres-Martínez, R.; Silva-Pereyra, H.G.; Reyes-Esqueda, J.A.; Rodríguez-Fernández, L.; Crespo-Sosa, A.; Cheang-Wong, J.C.; Oliver, A. Ultrafast optical phase modulation with metallic nanoparticles in ion-implanted bilayer silica. *Nanotechnology* **2011**, *22*, 355710. [CrossRef] [PubMed]
24. Badr, Y.; Mahmoud, M.A. Effect of PVA surrounding medium on ZnSe nanoparticles: Size, optical, and electrical properties. *Spectrochim. Acta Part A* **2006**, *65*, 584–590. [CrossRef] [PubMed]
25. Eslamian, M. Excitation by acoustic vibration as an effective tool for improving the characteristics of the solution-processed coatings and thin films. *Prog. Org. Coat.* **2017**, *113*, 60–73. [CrossRef]
26. Tripathy, S.; Marty, R.; Lin, V.K.; Teo, S.L.; Ye, E.; Arbouet, A.; Saviot, L.; Girard, C.; Han, M.Y.; Mlayah, A. Acousto-Plasmonic and Surface-Enhanced Raman Scattering Properties of Coupled Gold Nanospheres/Nanodisk Trimers. *Nano Lett.* **2011**, *11*, 431–437. [CrossRef] [PubMed]
27. Collis, J.F.; Chakraborty, D.; Sader, J.E. Autonomous propulsion of nanorods trapped in an acoustic field. *J. Fluid Mech.* **2017**, *825*, 29–48. [CrossRef]
28. McGraw-Hill. *Concise Encyclopedia of Science & Technology*, 5th ed.; McGraw-Hill Professional Publishing: New York, NY, USA, 2005; p. 18.
29. Lerosey, G.; Fink, M. Acousto-optic imaging: Merging the best of two worlds. *Nat. Photonics* **2013**, *7*, 265–267. [CrossRef]
30. Laude, V.; Belkhir, A.; Alabiad, A.F.; Addouche, M.; Benchabane, S.; Khelif, A.; Baida, F.I. Extraordinary nonlinear transmission modulation in a doubly resonant acousto-optical structure. *Optica* **2017**, *4*, 1245–1250. [CrossRef]
31. Gunther, J.; Andersson-Engels, S. Review of current methods of acousto-optical tomography for biomedical applications. *Front. Optoelectron.* **2017**, *10*, 211–238. [CrossRef]
32. Kauranen, M.; Zayats, A.V. Nonlinear plasmonics. *Nat. Photonics* **2012**, *6*, 737–748. [CrossRef]
33. Panoiu, N.C.; Sha, W.E.I.; Lei, D.Y.; Li, G.-C. Nonlinear optics in plasmonic nanostructures. *J. Optic.* **2018**, *20*, 083001. [CrossRef]
34. Valley, D.T.; Ferry, V.E.; Flannigan, D.J. Imaging Intra- and Interparticle Acousto-plasmonic Vibrational Dynamics with Ultrafast Electron Microscopy. *Nano Lett.* **2016**, *16*, 7302–7308. [CrossRef] [PubMed]
35. O'Brien, K.; Lanzillotti-Kimura, N.D.; Rho, J.; Suchowski, H.; Yin, X.; Zhang, X. Ultrafast acousto-plasmonic control and sensing in complex nanostructures. *Nat. Commun.* **2014**, *5*, 5042. [CrossRef] [PubMed]
36. Fabelinskii, I.L. *Molecular Scattering of Light*, 1st ed.; Springer: Berlin, Germany, 1968; pp. 81–85.
37. Yakovlev, V.V.; Dickson, W.; Murphy, A.; McPhillips, J.; Pollard, R.J.; Podolskiy, V.A.; Zayats, A.V. Ultrasensitive Non-Resonant Detection of Ultrasound with Plasmonic Metamaterials. *Adv. Mater.* **2013**, *25*, 2351–2356. [CrossRef] [PubMed]
38. Ou, J.-Y.; Plum, E.; Zhang, J.; Zheludev, N.I. Giant Nonlinearity of an Optically Reconfigurable Plasmonic Metamaterial. *Adv. Mater.* **2016**, *28*, 729–733. [CrossRef]
39. Chau, Y.-F.C.; Wang, C.-K.; Shen, L.; Lim, C.M.; Chiang, H.-P.; Chao, C.-T.C.; Huang, H.J.; Lin, C.-T.; Kumara, N.T.R.N.; Voo, N.Y. Simultaneous realization of high sensing sensitivity and tunability in plasmonic nanostructures arrays. *Sci. Rep.* **2017**, *7*, 16817. [CrossRef] [PubMed]
40. Bigall, N.C.; Parak, J.P.; Dorfs, D. Fluorescent, magnetic and plasmonic—Hybrid multifunctional colloidal nano objects. *Nano Today* **2012**, *7*, 282–296. [CrossRef]
41. Dowran, M.; Kumar, A.; Lawrie, B.J.; Pooser, R.C.; Marino, A.M. Quantum-enhanced plasmonic sensing. *Optica* **2018**, *5*, 628–633. [CrossRef]
42. Ahmed, D.; Peng, X.; Ozcelik, A.; Zheng, Y.; Huang, T.J. Acousto-plasmofluidics: Acoustic modulation of surface plasmon resonance in microfluidic systems. *AIP Adv.* **2015**, *5*, 097161. [CrossRef] [PubMed]

43. Hurtado-Aviles, E.A.; Torres, J.A.; Trejo-Valdez, M.; Urriolagoitia-Sosa, G.; Villalpando, I.; Torres-Torres, C. Acousto-Plasmonic Sensing Assisted by Nonlinear Optical Interactions in Bimetallic Au-Pt Nanoparticles. *Micromachines* **2017**, *8*, 321. [CrossRef] [PubMed]
44. Sonawane, R.S.; Dongare, M.K. Sol–gel synthesis of Au/TiO$_2$ thin films for photocatalytic degradation of phenol in sunlight. *J. Mol. Catal. Chem.* **2006**, *243*, 68–79. [CrossRef]
45. Boyd, R.W. *Nonlinear Optics*, 3rd ed.; Academic Press: New York, NY, USA, 2008.
46. Moniri, S.; Reza Hantehzadeh, M.; Ghoranneviss, M.; Asadi Asadabad, M. Au-Pt alloy nanoparticles obtained by nanosecond laser irradiation of gold and platinum bulk targets in an ethylene glycol solution. *Eur. Phys. J. Plus* **2017**, *132*, 318. [CrossRef]
47. Morales-Bonilla, S.; Torres-Torres, C.; Trejo-Valdez, M.; Torres-Torres, D.; Urriolagoitia-Sosa, G.; Hernández-Gómez, L.H.; Urriolagoitia-Calderón, G. Engineering the optical and mechanical properties exhibited by a titanium dioxide thin film with Au nanoparticles. *Opt. Appl.* **2013**, *43*, 651–661. [CrossRef]
48. Karkare, M.M. Choice of precursor not affecting the size of anatase TiO$_2$ nanoparticles but affecting morphology under broader view. *Int. Nano Lett.* **2014**, *4*, 111–118. [CrossRef]
49. Miguez, M.L.; De Souza, T.G.B.; Barbano, E.C.; Zilio, S.C.; Misoguti, L. Measurement of third-order nonlinearities in selected solvents as a function of the pulse width. *Opt. Express* **2017**, *25*, 3553–3565. [CrossRef] [PubMed]
50. Lea-Banks, H.; Teo, B.; Stride, E.; Coussios, C.C. The effect of particle density on ultrasound-mediated transport of nanoparticles. *Phys. Med. Biol.* **2016**, *61*, 7906. [CrossRef] [PubMed]
51. Quan, C.; He, M.; He, C.; Huang, Y.; Zhu, L.; Yao, Z.; Xu, X.; Lu, C.; Xu, X. Transition from saturable absorption to reverse saturable absorption in MoTe$_2$ nano-films with thickness and pump intensity. *Appl. Surf. Sci.* **2018**, *457*, 115–120. [CrossRef]
52. Torre, R.; Sánta, I.; Righini, R. Pre-transitional effects in the liquid-plastic phase transition of p-terphenyl. *Chem. Phys. Lett.* **1993**, *212*, 90–95. [CrossRef]
53. Fernández-Valdés, D.; Torres-Torres, C.; Martínez-González, C.L.; Trejo-Valdez, M.; Hernández-Gómez, L.H.; Torres-Martínez, R. Gyroscopic behavior exhibited by the optical Kerr effect in bimetallic Au-Pt nanoparticles suspended in ethanol. *J. Nanopart. Res.* **2016**, *17*, 204. [CrossRef]
54. Yesilkoy, F.; Terborg, R.A.; Pello, J.; Belushkin, A.A.; Jahani, Y.; Pruneri, V.; Altug, H. Phase-sensitive plasmonic biosensor using a portable and large field-of-view interferometric microarray imager. *Light Sci. Appl.* **2018**, *7*, 17152. [CrossRef] [PubMed]
55. Verdan, S.; Burato, G.; Coment, M.; Reinert, L.; Fuzellier, H. Structural changes of metallic surfaces induced by ultrasound. *Ultrason. Sonochem.* **2003**, *10*, 291–295. [CrossRef]
56. Kollath, A.; Cherepanov, P.V.; Andreeva, D.V. Controllable manipulation of crystallinity and morphology of aluminium surfaces using highintensity ultrasound. *Appl. Acoust.* **2016**, *103*, 190–194. [CrossRef]
57. Cheng, L.J.; Guo, L.J. Nanofluidic diodes. *Chem. Soc. Rev.* **2010**, *39*, 923–938. [CrossRef]
58. Yan, Y.; Warren, S.C.; Fuller, P.; Grzybowski, B.A. Chemoelectronic circuits based on metal nanoparticles. *Nat. Nanotechnol.* **2016**, *11*, 603–608. [CrossRef] [PubMed]
59. Lee, J.U.; Gipp, P.P.; Heller, C.M. Carbon nanotube p-n junction diodes. *Appl. Phys. Lett.* **2004**, *85*, 145–147. [CrossRef]
60. Gulbahar, B.; Memisoglu, G. Graphene-based Acousto-optic Sensors with Vibrating Resonance Energy Transfer and Applications. In *Two-Dimensional Materials for Photodetector*; IntechOpen Book: London, UK, 2018; pp. 179–192.
61. Becerril, D.; Batiz, H.; Pirruccio, G.; Noguez, C. Efficient Coupling to Plasmonic Multipole Resonances by Using a Multipolar Incident Field. *ACS Photonics* **2018**, *5*, 1404–1411. [CrossRef]
62. Tan, T.; Xie, H.; Xie, J.; Ping, H.; Su, B.-L.; Wang, W.; Wang, H.; Munir, Z.A.; Fu, Z. Photo-Assisted Synthesis of Au@PtAu Core-Shell Nanoparticles with Controllable Surface Composition for Methanol Electrooxidation. *J. Mater. Chem. A* **2016**, *4*, 18983–18989. [CrossRef]
63. Gowthaman, N.S.K.; Sinduja, B.; Shankar, S.; John, S.A. Displacement Reduction Routed Au-Pt Bimetallic Nanoparticles: A Highly Durable Electrocatalyst for Methanol Oxidation and Oxygen Reduction. *Sustain. Energy Fuels* **2018**, *2*, 1588–1599. [CrossRef]
64. Mesch, M.; Metzger, B.; Hentschel, M.; Giessen, H. Nonlinear Plasmonic Sensing. *Nano Lett.* **2016**, *16*, 3155–3159. [CrossRef]

65. Gérard, D.; Laude, V.; Sadani, B.; Khelif, A.; Labeke, D.V.; Guizal, B. Modulation of the extraordinary optical transmission by surface acoustic wave. *Phys. Rev. B* **2007**, *76*, 235427. [CrossRef]
66. Ganeev, R.A.; Ryasnyanskii, A.I.; Kodirov, M.K.; Kamalov, S.R.; Usmanov, T. Nonlinear Optical Characteristics of Colloidal Solutions of Metals. *Opt. Spectrosc.* **2001**, *90*, 568–573. [CrossRef]
67. Kedenburg, S.; Vieweg, M.; Gissibl, T.; Giessen, H. Linear refractive index and absorption measurements of nonlinear optical liquids in the visible and near-infrared spectral region. *Opt. Mater. Express* **2012**, *2*, 1588–1611. [CrossRef]
68. Pei, Y.; Yao, F.; Ni, P.; Sun, X. Refractive index of silver nanoparticles dispersed in polyvinyl pyrrolidone nanocomposite. *J. Mod. Opt.* **2010**, *57*, 872–875. [CrossRef]

© 2019 by the authors. Licensee MDPI, Basel, Switzerland. This article is an open access article distributed under the terms and conditions of the Creative Commons Attribution (CC BY) license (http://creativecommons.org/licenses/by/4.0/).

Article

Holographic Performance of Azo-Carbazole Dye-Doped UP Resin Films Using a Dyeing Process

Kenji Kinashi [1,*], Isana Ozeki [2], Ikumi Nakanishi [2], Wataru Sakai [1] and Naoto Tsutsumi [1,*]

1. Faculty of Materials Science and Engineering, Kyoto Institute of Technology, Kyoto 606-8585, Japan; wsakai@kit.ac.jp
2. Master's Program of Innovative Materials, Graduate School of Science and Technology, Kyoto Institute of Technology, Matsugasaki, Sakyo, Kyoto 606-8585, Japan; isana.oz.0611@gmail.com (I.O.); m7616023@edu.kit.ac.jp (I.N.)
* Correspondence: kinashi@kit.ac.jp (K.K.); tsutsumi@kit.ac.jp (N.T.)

Received: 25 February 2019; Accepted: 12 March 2019; Published: 21 March 2019

Abstract: For the practical application of dynamic holography using updatable dyed materials, optical transparency and an enlarged sample size with a uniform dispersion of the dye and no air bubbles are crucial. The holographic films were prepared by applying a dyeing method comprising application, curing, dyeing, and washing to an unsaturated polyester (UP) resin film. The unsaturated polyester (UP) resin film with high optical transparency was dyed with a 3-[(4-cyanophenyl)azo]-9H-carbazole-9-ethanol (CACzE) (azo-carbazole) dye via the surfactant, polyoxyethylene (5) docosyl ether, in an aqueous solution. The amount of dye uptake obtained via the dyeing process ranged from 0.49 to 6.75 wt.%. The dye concentration in the UP resin was proportional to the dye concentration in the aqueous solution and the immersion time. The UP resin film with 3.65 wt.% dye exhibited the optical diffraction property η_1 of 0.23% with a response time τ of 5.9 s and a decay time of 3.6 s. The spectroscopic evaluation of the UP resin film crosslinking reaction and the dyeing state in the UP resin film are discussed. Furthermore, as an example of its functionality, the dynamic holographic properties of the dye-doped UP resin film are discussed.

Keywords: dyeing; unsaturated polyester resin; azobenzene; hologram; aqueous dispersion

1. Introduction

Polymer-based dynamic holographic films have attracted considerable attention as a next-generation 3D display technology [1–7]. Optical transparency and an enlarged sample size with a uniform dispersion of the dye and no air bubbles are the minimum requirements for the practical application of dynamic holograms [3]. In our 2016 report [3], we proposed insightful scientific research to allow transparency in the visible region for the azo-carbazole analog based on spectroscopic and holographic optical perspectives. Melt-pressed poly(methyl methacrylate) (PMMA) films dispersed with azo-carbazole dyes give high optical transparency with a uniform coloring state and no air bubbles. Making an enlarged size film via melt-press on a laboratory scale, however, has some limitations. Notably, the fabrication of the large sample size would be challenging. Therefore, the development of a new method to replace the melt-press method is necessary.

Dye-doped polymeric films possess well-known properties, such as the manipulation and detection of light. Therefore, these films have been used as optoelectronic devices, such as organic dye lasers, organic light-emitting diodes (LED), dye photodetectors, dye-sensitized solar cells, and displays [8–11]. Solution processes such as spin-coating and dip-coating are common methods for dispersing dyes into a polymer matrix due to their convenience and safety [12,13]. However, these solution processes cause dye aggregation or phase separation during or after drying that can lead to serious defects in the electro-optical properties [14]. The uniform dispersion of dyes and pigments into a polymer matrix is a key process for the fabrication of optoelectronic devices.

Dyestuffs are substances that have the dyeing capacity of fibers and films. When the dyestuffs and fibers or films have a chemical affinity for each other, the dyestuffs can easily penetrate the fibers or films. Dyeing is a useful tool to uniformly color fibers and films. The coloring process, in which the dyes are allowed to uniformly penetrate fibers and films, is performed by immersing the dyes in a liquid solution dispersed with dyestuffs. Since the fibers and films are colored by this technique to impregnate the dye itself with a dye, there is no interlayer peeling at the interface, as with a coating. Therefore, a stabilized functional fiber without dye aggregation or phase separation can possibly be obtained.

Highly transparent films can be prepared from unsaturated polyester (UP) resins. The application process of the UP resins before curing can be accomplished in a number of convenient processes such as spraying, deposition with an applicator, dipping, or melt-press.

In this study, we focused on the dyeing process using aqueous dispersion; then, we introduced a typical azo-carbazole dye, CACzE, to large size UP resin films, and, finally, evaluated the dyeing properties of the dye-doped UP resin films and their dynamic holographic properties.

2. Experimental Section

The 3-[(4-cyanophenyl)azo]-9*H*-carbazole-9-ethanol (CACzE) was synthesized according to a previously reported procedure [3]. Figure 1 shows the structural formula of CACzE and a photograph of the CACzE powder. The UP resin, as a host polymeric resin, and a curing agent, Permec N, which contains ethyl methyl ketone peroxides, dimethyl phthalate, and ethyl methyl ketone, were purchased from FRP-ZONE Co., Saitama, Japan. The surfactant polyoxyethylene (5) docosyl ether was purchased from Wako Co., Osaka, Japan. The mixtures containing 98% UP resin and 2% curing agent were vigorously stirred and deposited on a glass substrate by using a coating applicator with a 100 μm gap. The aqueous dispersion dyebaths were made via the dispersion of CACzE portions into 100 g of distilled water to provide 1.9×10^{-4} to 7.6×10^{-3} mol L^{-1} solutions (referred to herein as the aqueous dispersion). Additionally, specific amounts of the surfactant were added to the aqueous dispersion to adsorb the dispersed CACzE into the UP resin after curing. A summary of the aqueous dispersions is listed in Table 1. During the aqueous dispersion treatment process, the cured UP resins were immersed in the aqueous dispersion and shaken at 120 °C for various immersion cycles. The UP resins dyed with the aqueous dispersion were washed with dimethyl sulfoxide (DMSO) and air dried. Schematic diagrams of each process are shown in Figure 2.

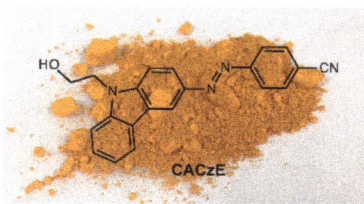

Figure 1. Structural formula of CACzE and background photograph of CACzE powder.

The concentration of CACzE in the UP resin was recorded on a spectrophotometer (Lambda 1050 UV/Vis/NIR, Perkin Elmer, Waltham, MA, USA) and a laser Raman microscope (Raman-11, Nanophoton, Osaka, Japan) at room temperature. The Raman spectra were recorded under a 785 nm laser excitation. Transmittance spectra of the UP resin during the crosslinking reaction were measured by a Fourier Transform-Infrared Spectroscopy (FT-IR) spectrophotometer (FT/IR-4700 with ATR PRO ONE equipped with a diamond prism, Jasco, Tokyo, Japan). The ATR FT-IR spectrophotometer with a resolution of 1 cm^{-1} in the transmission mode was used for kinetic measurement of the crosslinking reaction. The glass-transition temperature (T_g) was determined by a differential scanning calorimetry (DSC) (DSC2920, TA Instruments Co., New Castle, DE, USA) at a heating rate of 10 °C min^{-1}. The haze value was measured to evaluate the transparency and scattering properties of the sample films using an integrating sphere. The haze value (%) was measured using an integral sphere and calculated as the total light intensity of the scattered light divided by the total light intensity of the sum of scattered and transmitted light. The geometry of the haze value measurement system is shown in Figure S1. The external diffraction efficiency was measured using a 4f reduction projection system, as shown in Figure 3. A vertical fringe pattern image with an s-polarized 532 nm writing beam reflected through a polarizing beam-splitter was projected directly onto the sample. A PC-controlled vertical fringe interference pattern with a grating number of 100 on a special light modulator (SLM) (1920 pixels wide × 1080 pixel high; 8.0 μm pixel size, HOLOEYE Photonics Co., Pittsfield, MA, Germany) provided a fringe pattern spacing of Λ = 25 μm on the sample surface. A weak s-polarized reading beam of 1 mW with a DPSS laser at 640 nm (Bolero^{TM}, Cobolt Co., Solna, Sweden) was illuminated on the sample surface, and the first-order diffraction intensity from the resultant refractive index gratings was measured by a silicon photodiode.

Table 1. Experimental conditions for the aqueous dispersions.

Entry	CACzE Aqueous Dispersion (10^{-4} mol L^{-1})	Surfactant (g)
1	1.9	0.1
2	3.8	0.2
3	5.7	0.3
4	7.6	0.4
5	9.5	0.5
6	11.4	0.6
7	13.3	0.7
8	15.2	0.8
9	17.1	0.9
10	19.0	1.0
11	38.0	2.0
12	76.0	4.0

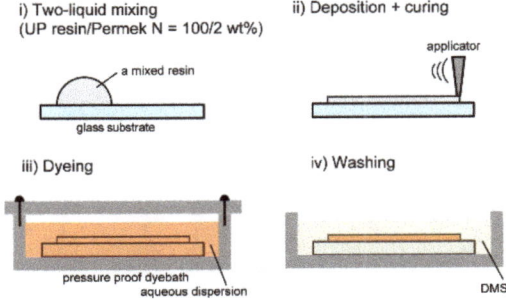

Figure 2. Schematic of the experimental procedure for the dispersion dyeing method. (i) A two-liquid mixing resin (UP resin/Permek N = 100/2 wt.%) deposited on a glass substrate; (ii) coating with an applicator with a 100 μm gap and curing; (iii) dyeing in the pressure proof dyebath; and (iv) washing with DMSO.

We used the intensities of incident light (I_0) and first-order diffracted light (I_{d1}) to evaluate the first-order diffraction efficiency η_1 with Equation (1):

$$\eta_1 = \frac{I_{d1}}{I_0} \times 100. \tag{1}$$

The response time τ of the first-order diffraction efficiency η_1 as a function of time was fitted by the Kohlrausch-Williams-Watts stretched exponential function in Equation (2),

$$\eta_1 = \eta_0 \left\{ 1 - \exp\left[-\left(\frac{t}{\tau}\right)^\beta \right] \right\}, \qquad (2)$$

where t is the time, η_0 is the steady-state external diffraction efficiency, and β ($0 < \beta \leq 1$) is the parameter related to a dispersion.

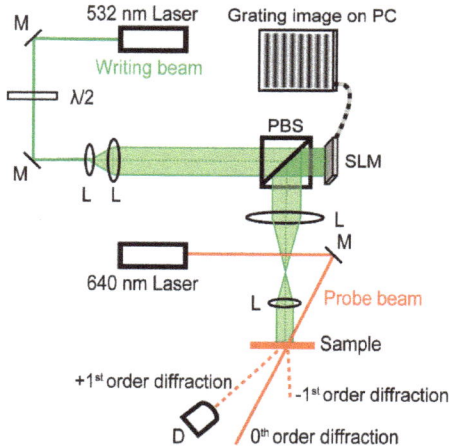

Figure 3. Schematic representation of the 4f reduction projection system: L—lens, M—mirror; PBS—polarizing beam splitter, PC—personal computer SLM—spatial light modulator, and D—photodiode. Laser sources are a green CW laser at 532 nm for recording and a red laser at 640 nm for reading.

3. Results and Discussion

The UP resin was synthesized using the free radical chain-growth crosslinking reaction of an unsaturated polyester and styrene monomer with curing reagents, as shown in Figure 4a. During the crosslinking reaction at room temperature, the peroxides in the curing reagents acted as catalysts. The qualitative degree of the crosslinking in the UP resin containing 2 wt.% curing reagents was investigated using the ATR-FT/IR absorption measurements and analysis. Figure 4b shows the ATR-FT/IR absorption spectra for the UP resin after the addition of the curing reagents in the film. These spectra were measured in the dark, over time, at room temperature, and their spectra at each elapsed time were used for the kinetic evaluation of the crosslinking reaction. The ATR FT-IR spectrum of the as-prepared UP resin film shows absorptions at 3081, 3059, 3026, 2982, 2955, 1728, 1646, 1630, 1600, 1579, 1494, 1449, 1371, 1286, 1127, 1071,1042, 1021, 992, 910, 846, 778, 743, and 701 cm^{-1}. These bands are ascribed to the terminal methylene C–H stretching (3081 cm^{-1}), aromatic C–H stretching (3059 and 3026 cm^{-1}), symmetric C–H stretching 2982 cm^{-1}, aliphatic C–H stretching (2955 cm^{-1}), C=O stretching (1728 cm^{-1}), C=C stretching (1646 and 1630 cm^{-1}), aromatic C=C stretching (1600 and 1579 cm^{-1}), C–H bending (1494 cm^{-1}), C–H bending (1449 cm^{-1}), aliphatic C–H stretching in methyl (1371 cm^{-1}), Ph–C=O stretching (1286 cm^{-1}), Ph–C–O stretching (1127 cm^{-1}), C–H in-plane deformation (1071 cm^{-1}), out-of-plane C–H bending in CH=CH$_2$ (992 and 910 cm^{-1}), C–O–C stretching (846 cm^{-1}), C=C stretching in CH=CH$_2$ (778 cm^{-1}), and aromatic out-of-plane C–H bending (743 and 701 cm^{-1}) of the UP resin and styrene. The transmittances based on the absorption of C=C stretching at 1646 and 910 cm^{-1}, out-of-plane C–H bending in CH=CH$_2$ at 992 and 910 cm^{-1}, and C=C stretching in CH=CH$_2$ at 778 cm^{-1} are significantly increased upon increase of the curing time, implying the reduction of the C=C bonds is due to the progress of the crosslinking reaction at room temperature. The reduction of the C=C in styrene during the crosslinking reaction was determined by the transmittance change, $\Delta T/T_0$ (ΔT is an absolute change in the transmittance given by $T-T_0$, where T is the transmittance at a time after curing and T_0 is the transmittance before curing), at 778 cm^{-1}. The plots of the transmittance change $\Delta T/T_0$ leveled out at approximately 24 h curing, which indicated that the crosslinking reaction was almost complete at 24 h. It should be noted here that most of the styrene monomer does not remain in the UP resin film after curing for 24 h. Because the reaction between the styrene monomer and fumaric acid ester is faster than the reaction between styrene monomers, it is assumed that styrene polymer does not remain in the UP resin film. As shown in Figure S2, the glass-transition temperature (~100 °C.) of the styrene polymer was not detected. The glass-transition temperature, T_g, of the UP resin film after curing for 24 h was 33.2 °C. This glass-transition temperature of the UP resin film indicated that the amount of styrene contained in the UP would be 6% or less [15]; it has been reported that the UP resin is a stable film without weight loss up to about 300 °C. In addition, swelling and shrinkage before and after curing of the UP resin were not confirmed. Descriptions of the complicated curing mechanisms of the UP resin have been extensively reported in the literature [16]. Optical transparency is one of the most important characteristics of the holographic display. The haze value (%), defined by the ratio of diffuse transmittance (%)/total light transmittance (%) × 100, was measured for the UP resin film after curing and before dyeing, and was found to be 5.0%. Although a haze value of 4% or below is prefereable for holographic display applications, the haze values are acceptable for reflection and transmission holography. Furthermore, an UP resin film with a haze value of 5% is a smooth surface and shows high transparency with no scattering; this value is similar to the polymethyl methacrylate (PMMA) haze value of 2.6% [7]. The evaluation of the haze value for the obtained UP resin film is useful for the performance of holograms and other research fields, such as organic light emitting devices.

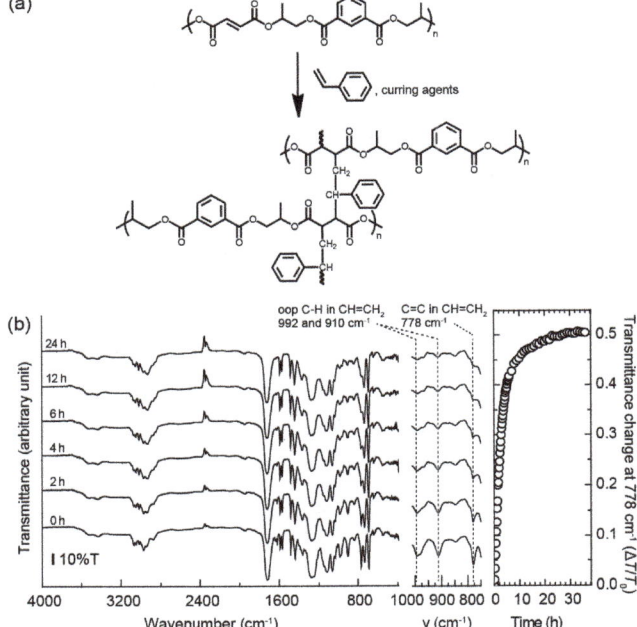

Figure 4. (a) Crosslinking reaction of the UP resin. (b) ATR-FT/IR spectra of a UP resin film and the transmittance change $\Delta T/T_0$ at 778 cm^{-1}.

The UP resin films are immersed in the aqueous dispersion dyebath and heated at a temperature above the T_g of the UP resin before the CACzE dyes are absorbed into the UP resin film. The CACzE dyes that remain at the surface of the UP resin films are removed with DMSO. The concentration of the CACzE dye uptake in the UP resin films was estimated based on the absorbance at 561 nm, as shown in Figure S3. The concentration of the CACzE dye uptake increased from 0.49 to 6.75 wt.% by increasing the concentration of the aqueous dispersion and the immersion time (all results are summarized in Table S1). The relationship between the equilibrium of CACzE dye uptake and the concentration of CACzE aqueous solution is shown in Figure 5a, and the Freundlich isotherm model was applied to the equilibrium data. For ordinary adsorption, the Langmuir adsorption isotherm model, showing a saturation point with increasing concentrations, should be adopted. However, it is suggested that the adsorption curves of the UP resin films appear not to be saturated; that is, the Freundlich isotherm model, which fits well with the adsorption behavior in the low concentration region, would be appropriate [17]. As a result, the maximum concentration of the CACzE dye uptake in the UP resin film was 6.75 wt.% at 7.60×10^{-3} mol L^{-1} CACzE aqueous dispersion (Entry 12), with a dyebath temperature at 120 °C, and an immersion time of 12 h. When the concentration, immersion time or dyebath temperature were higher than these conditions, which demonstrated the maximum dye uptake, the UP resin film resulted in more light scattering and brittle texture. Figure 5b shows a photograph of a 100 mm square size UP resin film with a dye uptake of 3.65 wt.% after dyeing (Entry 10, immersion time 12 h). The large size UP resin film was uniformly dyed, and its transparency was remained relatively high, showing a smooth surface; however, the haze value was 11.6%. Figure 5c shows the depth profiles of the dye-doped UP resin film (Entry 10), where the Raman peak of Ph–N= [18] unit in CACzE dye at 1148 cm^{-1} was measured using laser Raman spectroscopy, and are plotted as a function of the measurement depth. Figure 5d shows a schematic illustration for the laser Raman measurement and the schematic profile of dye concentration distribution in the direction of film depth. The CACzE dye concentration vs. the depth profile shows a diffusion-controlled distribution of the CACzE dye concentration, which indicates that the CACzE concentration declines as the depth into the film bulk increases. The average concentration of the CACzE dye uptake in the UP resin film dyed at 1.90×10^{-3} mol L^{-1} CACzE aqueous dispersion (Entry 10), a dyebath temperature at 120 °C, and the immersion time for 12 h was 3.65 wt.% based on the absorbance at 561 nm. On the other hand, as shown in Figure 5c, the concentration of the CACzE dye in the range of 10 μm from the surface was 7.53 wt.%; however, the CACzE dye concentration significantly decreased at the deeper position in the film. This dyeing process is a noteworthy technique for fabricating a ~10 μm thick film with a dye concentration of 7 wt.% or higher, which can be much more powerful and effective for large size film than the spin-coating technique. Furthermore, the advantage of the dyeing process is that it is not a batch process and therefore has a relatively high throughput compared to the spin-coating process. Finally, the actual amount of dye used in the dyeing process was very low, and it was possible to incorporate all of dye into the substrate. The dyeing process will certainly be effective for holographic application requiring large area transparency. The depth profiles are noted; the CACzE concentration is near zero at a depth of 70 μm, and a film thickness of 70 μm or less is preferable in this dyeing process.

The holographic gratings for the films containing azobenzene moieties were induced by two kinds of processes; a modulation of the polarization grating due to the nanoscopic angular reorientation of azobenzene moieties and a modulation of the surface relief grating induced by a macroscopic molecular migration of azobenzene molecules. The latter is well-known to produce thermally stable gratings [19]. In the present case, no surface relief gratings were observed on the surface of the dye-doped UP resin films.

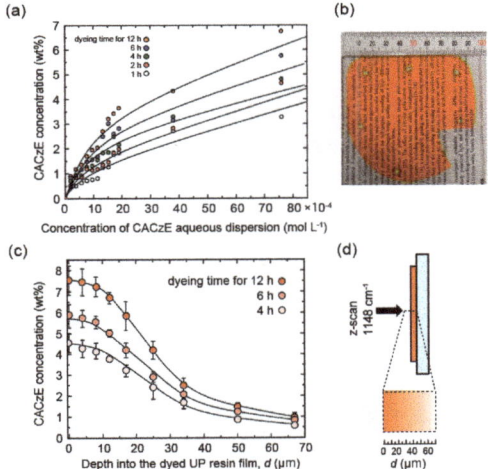

Figure 5. (a) Dye uptake isotherms of the CACzE dye on the UP resin film (dyebath temp. 120 °C); (b) a photograph of an UP resin film with 100 mm square after dyeing; (c) CACzE concentration vs. depth profiles for the dye-doped UP resin films (Entry 10); and (d) a schematic illustration of dye-doped UP resin film.

The first order diffraction efficiency, η_1, was plotted as a function of the recording time, followed by the elapsed time at room temperature for the dye-doped UP resin film (Figure 6). The interference fringe pattern of the writing beam was turned on at time zero. As the time elapsed, the increase in the first-order diffraction efficiency η_1 was measured. The writing beam of the interference fringe pattern was turned off at 50 s. As a result, a steady-state η_1 of 0.23%, a response (rising) time τ of 5.9 s, and a decay time of 3.6 s were measured. In comparison, a melt-pressed PMMA film containing CACzE of 3.65 wt.% (CACzE/PMMA), and a first-order diffraction efficiency η_1 at 5.9 s was almost the same value (η_1 = 0.55%); however, the haze value of the dye-doped UP resin film was slightly higher than the value of the CACzE/PMMA of 3.65 wt.%. Therefore, the optical loss seemed to be lowering the first-order diffraction efficiency of the dye-doped UP resin film. The diffraction grating can be classified as the Raman-Nath regime because the product of the grating thickness d and the writing beam wavelength λ in the film was smaller than the square of the fringe pattern spacing Λ, $\Lambda^2 > d\lambda$. The equation for the theoretical grating thickness d_t is $I = I_0 \exp(-\alpha d_t)$, where an absorption coefficient α = 126 cm^{-1} at 532 nm, and it is estimated to be 365 µm. The absorption coefficient α of the dyed-doped UP resin film was given by absorption spectral feature (Figure S3, Supplementary Materials). Accordingly, $\Lambda^2 > d\lambda$ held, and the diffraction grating could be determined as the Raman-Nath regime. The result indicated that it was clearly different from the azo-mesogenic polymers or azo-elastomers showing Bragg diffraction derived from the surface relief grating which has been reported so far [20–23] The refractive index modulation Δn in a transmitted Raman-Nath grating is proportional to the diffraction efficiency for sinusoidal phase grating expressed by first-order Bessel functions [24,25]:

$$\eta_1 = J_1^2(\delta) = J_1^2\left(\frac{2\pi d \Delta n}{\lambda \cos\theta}\right) \quad (3)$$

where η_1 is the first-order diffraction efficiency, J_1 is the first-order Bessel function, d is the grating thickness, Δn is the refractive index modulation, θ is the incidence angle of the reading beam within the film, and δ is the Raman-Nath parameter. Thus, a refractive index modulation Δn was estimated to be 1.3 × 10^{-4} with first-order diffraction efficiency of 0.23% and the other parameters were: θ = 21.5°, λ = 640 nm, δ = 0.096, and d = 70 µm. The grating thickness d considered here corresponds to the measured depth profile of the CACzE concentration. However, the refractive index modulation Δn of 1.3×10^{-4}, corresponding to the steady-state η_1 of 0.23%, showed a relatively low value compared to the values previously reported [7]. The cause of this low diffraction efficiency in the film is established in Figure 5c. The concentration of dyes had a significant gradient along the depth into the UP resin film. This concentration gradient of the CACzE dyes may be responsible for the low optical diffraction. Another possible reason is the difference of the matrix. The former matrix was poly(methyl methacrylate) (PMMA) and the present matrix is epoxy resin. The environment around the CACzE dyes in the matrix significantly affect the photo-isomerization process. For example, the interaction between the CACzE dyes and matrix and/or the free volume in the matrix allowing the rotation of the trans-cis photo-isomerization would affect the optical diffraction. These points should be clarified in future studies. However, the present results of the diffraction properties in the dye-doped UP resin film provide the success of the holographic film using the dyeing process.

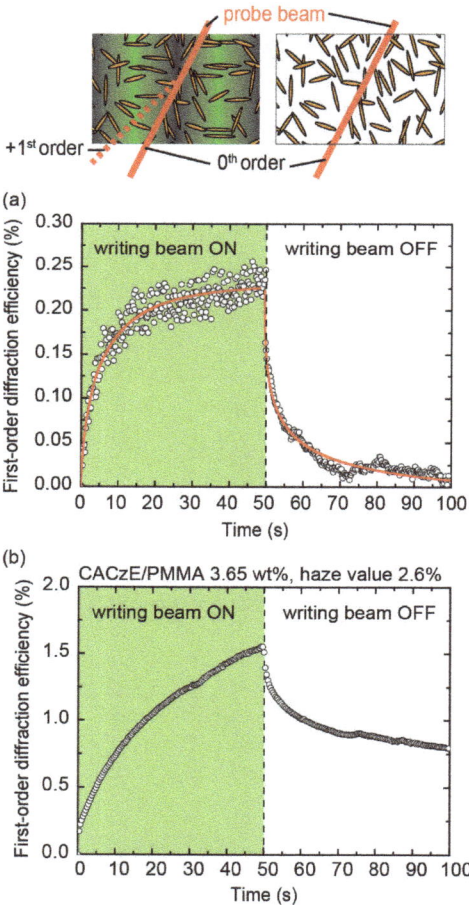

Figure 6. Top-view schematic representation of diffraction responses of probe beam for perpendicular orientation (left) and random orientation (right) of the CACzE dyes. (**a**) Time evolution of the first-order diffraction efficiency for the dye-doped UP resin film with 3.65 wt.% (Entry 10, immersion time 12 h). (**b**) Time evolution of the first-order diffraction efficiency for the melt-press CACzE/PMMA with 3.65 wt.%.

The response times of the dye-doped UP resin film show a faster response than those of the melt-pressed CACzE/PMMA film (the response time is not saturated within 50 s as shown in Figure 6b), which may be due to the higher glass-transition temperature T_g of the CACzE/PMMA film than the dye-doped UP resin film. The thermal decay of η_1 for the dye-doped UP resin film after being turned off is shown in Figure 6a. The decay time corresponded to the single exponential decay function, and the decay time constant τ_d was estimated to be 3.6 s, faster than that of the melt-pressed CACzE/PMMA film. The holographic properties of the dye-doped UP resin film obtained in the dyeing process were relatively low in comparison to a standard melt-pressed PMMA film containing 30 wt.% CACzE dye; however, the film size was 16 times larger than the standard film. In other words, if a large size, as well as high concentration, can be achieved by improving the dyeing process, a large size film with high holographic properties will be obtained.

In conclusion, the dyeing process successfully introduced the CACzE dyes into the UP resin film in an aqueous solution. The gradient concentration ranging from 6.75 to 0.49 wt.% was measured along the depth from the sample surface, and the concentration in the range of 10 µm from the surface was approximately twice as high as the average concentration of 3.65%. The typical holographic characteristics, including a steady-state η_1 of 0.23%, response time τ of 5.9 s, and decay time of 3.6 s, are given.

4. Conclusions

The holographic properties of a CACzE azo-carbazole dye in a UP resin were investigated. The crosslinking of the UP resin containing 2 wt.% curing agents was evaluated using an ATR-FT/IR analysis. The CACzE dyes were successfully dispersed into the UP resin film in the aqueous dye solution. The total amount of dye uptake during the dyeing process ranged from 0.49 to 6.75 wt.%. The dye-doped UP resin films using the dyeing process showed that the resulting concentrations of the CACzE dye uptake increased in proportion to the concentration of the aqueous solution and the immersion time, ranging from 0.49 to 6.75 wt.%. The dye-doped UP resin film with 3.65 wt.% exhibited a steady-state holographic diffraction efficiency of η_1 of 0.23%, response time of τ of 5.9 s, and decay time of 3.6 s. The present dyeing process using aqueous solutions is a contribution to, and an advantage for, fabricating large-sized holographic devices as well as fabricating the photonic devices based on any polymer film containing organic dye.

Supplementary Materials: The following are available online at http://www.mdpi.com/1996-1944/12/6/945/s1, Figure S1: Haze value measurement system. (a) Measuring total transmitted light intensity. (b) Measuring scattered transmitted light intensity. A collimated light of 636 nm was used as the probe beam. Figure S2: DSC thermogram with heat flow signal vs. temperature for the UP resin film after curing for 24 h. Figure S3: UV–visible absorption spectra of the dyed UP resin films at each immersion time. Table S1: Film thickness, absorbance, dye uptake for the UP resin films after dyeing processes.

Author Contributions: Conceptualization, K.K.; methodology, K.K.; validation, K.K., W.S. and N.T.; investigation, I.O. and I.N.; Supervision, K.K. and N.T.; writing—original draft preparation, K.K.; writing—review and editing, N.T.; visualization, K.K.; funding acquisition, N.T.

Funding: This research was funded by the Program for Strategic Promotion of Innovative Research and Development (S-Innovation), Japan Science and Technology Agency (JST).

Conflicts of Interest: The authors declare no conflicts of interest.

References

1. Davidenko, N.A.; Davidenko, I.I.; Pavlov, V.A.; Tarasenko, V.V. Experimental investigations of the relaxation of polarization holograms in films of azobenzene polymers with chromophores with different substitutes. *Optik* **2018**, *165*, 174–178. [CrossRef]
2. Tsutsumi, N. Recent advances in photorefractive and photoactive polymers for holographic applications. *Polym. Int.* **2017**, *66*, 167–174. [CrossRef]

3. Kinashi, K.; Fukami, T.; Yabuhara, Y.; Motoishi, S.; Sakai, W.; Kawamoto, M.; Sassa, T.; Tsutsumi, N. Molecular design of azo-carbazole monolithic dyes for updatable full-color holograms. *NPG Asia. Mater.* **2016**, *8*, e311. [CrossRef]
4. Mao, W.; Sun, Q.; Baig, S.; Lu, H.; Wang, M.R. Red light holographic recording and readout on an azobenzene–LC polymer hybrid composite system. *Opt. Commun.* **2015**, *355*, 256–260. [CrossRef]
5. Tsutsumi, N.; Kinashi, K.; Tada, K.; Fukuzawa, K.; Kawabe, Y. Fully updatable three-dimensional holographic stereogram display device based on organic monolithic compound. *Opt. Express* **2013**, *21*, 19880–19884. [CrossRef]
6. Tsutsumi, N.; Kinashi, K.; Sakai, W.; Nishide, J.; Kawabe, Y.; Sasabe, H. Real-time three-dimensional holographic display using a monolithic organic compound dispersed film. *Opt. Mater. Express* **2012**, *2*, 1003–1010. [CrossRef]
7. Kinashi, K.; Nakanishi, I.; Sakai, W.; Tsutsumi, N. Material Design of Azo-Carbazole Copolymers for Preservation Stability with Rewritable Holographic Stereograms. *Macromol. Chem. Phys.* **2018**, *220*, 1800456. [CrossRef]
8. Zhang, L.; Li, C.; Li, J.; Jiang, F. Controlling directions of electron flow by light: A case study on TiOfilm with azo dyes. *Dyes Pigment.* **2019**, *161*, 277–282. [CrossRef]
9. Kido, J.; Shinoyama, H.; Nagai, K. Single-layer white light-emitting organic electroluminescent devices based on dye-dispersed poly(N-vinylcarbazole). *Appl. Phys. Lett.* **1995**, *67*, 2281. [CrossRef]
10. Aihara, S.; Hirano, Y.; Tanioka, K.; Abe, M.; Saito, N.; Kamata, N.; Terunuma, D. Wavelength selectivities of organic photoconductive films: Dye-doped polysilanes and zinc phthalocyanine/tris-8-hydroxyquinoline aluminum double layer. *Appl. Phys. Lett.* **2003**, *82*, 511. [CrossRef]
11. Imoto, K.; Takahashi, K.; Yamaguchi, T.; Komura, T.; Nakamura, J.; Murata, K. Merocyanine Dye-Sensitization of Polythiophene in a Conjugated Polymer/TiOp–n Hetero-Junction Solar Cell. *Bull. Chem. Soc. Jpn.* **2003**, *76*, 2277–2283. [CrossRef]
12. Benamar, E.; Rami, M.; Messaudi, C.; Sayah, D.; Ennaoui, A. Structural, optical and electrical properties of indium tin oxide thin films prepared by spray pyrolysis. *Solar Energy Mater. Solar Cells* **1998**, *56*, 125–139. [CrossRef]
13. Shakti, N. Structural and Optical Properties of Sol-gel Prepared ZnO thin film. *Appl. Phys. Res.* **2001**, *2*, 19–28. [CrossRef]
14. Tanigaki, N.; Mochizuki, H.; Mo, X.; Mizokuro, T.; Hiraga, T.; Taima, T.; Yase, K. Molecular doping of poly(p-phenylenevinylene) under vacuum for photovoltaic application. *Thin Solid Films* **2006**, *499*, 110–113. [CrossRef]
15. Sanchez, E.M.S.; Zavaglia, C.A.C.; Felisberti, M.I. Unsaturated polyester resins: Influence of the styrene concentration on the miscibility and mechanical properties. *Polymer* **2000**, *41*, 765–769. [CrossRef]
16. Cao, X.; Lee, L.J. Control of shrinkage and residual styrene of unsaturated polyester resins cured at low temperatures: I. Effect of curing agents. *Polymer* **2003**, *44*, 1893–1902. [CrossRef]
17. Daneshvar, M.; Hosseini, M.R. Kinetics, isotherm, and optimization of the hexavalent chromium removal from aqueous solution by a magnetic nanobiosorbent. *Environ. Sci. Pollut. Res. Int.* **2018**, *25*, 28654–28666. [CrossRef]
18. Wu, Y.; Zhao, B.; Xu, W.; Li, B.; Jung, Y.M.; Ozaki, Y. Near-Infrared Surface-Enhanced Raman Scattering Study of Ultrathin Films of Azobenzene-Containing Long-Chain Fatty Acids on a Silver Surface Prepared by Silver Mirror and Nitric Acid Etched Silver Foil Methods. *Langmuir* **1999**, *15*, 4625–4629. [CrossRef]
19. Koskela, J.E.; Vapaavuori, J.; Hautala, J.; Priimagi, A.; Faul, C.F.J.; Kaivola, M.; Ras, R.H.A. Surface-Relief Gratings and Stable Birefringence Inscribed Using Light of Broad Spectral Range in Supramolecular Polymer-Bisazobenzene Complexes. *J. Phys. Chem. C* **2012**, *116*, 2363–2370. [CrossRef]
20. Ho, T.-J.; Chen, C.-W.; Khoo, I.C. Polarisation-free and high-resolution holographic grating recording and optical phase conjugation with azo-dye doped blue-phase liquid crystals. *Liq. Cryst.* **2018**, *45*, 1944–1952. [CrossRef]
21. Ciuchi, F.; Mazzulla, A.; Cipparrone, G. Permanent polarization gratings in elastomer azo-dye systems: Comparison of layered and mixed samples. *JOSA B* **2002**, *19*, 2231–2537. [CrossRef]
22. Ciuchi, F.; Mazzulla, A.; Carbone, G.; Cipparrone, G. Complex Structures of Surface Relief Induced by Holographic Recording in Azo-Dye-Doped Elastomer Thin Films. *Macromolecules* **2003**, *36*, 5689–5693. [CrossRef]

23. Sio, L.D.; Serak, S.; Tabiryan, N.; Umeton, C. Mesogenic versus non-mesogenic azo dye confined in a soft-matter template for realization of optically switchable diffraction gratings. *J. Mater. Chem. C* **2011**, *21*, 6811–6914. [CrossRef]
24. Goodman, J.W. An introduction to the principles and applications of holography. *Proc. IEEE* **1971**, *59*, 1292. [CrossRef]
25. Eichler, H.J.; Massmann, F.; Biselli, E.; Richter, K.; Glotz, M.; Konetzke, L.; Yang, X. Laser-induced free-carrier and temperature gratings in silicon. *Phys. Rev. B* **1987**, *36*, 3247–3253. [CrossRef]

© 2019 by the authors. Licensee MDPI, Basel, Switzerland. This article is an open access article distributed under the terms and conditions of the Creative Commons Attribution (CC BY) license (http://creativecommons.org/licenses/by/4.0/).

Article
Shape-Memory Assisted Scratch-Healing of Transparent Thiol-Ene Coatings

Algirdas Lazauskas *, Dalius Jucius, Valentinas Baltrušaitis, Rimantas Gudaitis, Igoris Prosyčevas, Brigita Abakevičienė, Asta Guobienė, Mindaugas Andrulevičius and Viktoras Grigaliūnas

Institute of Materials Science, Kaunas University of Technology, K. Baršausko 59, LT51423 Kaunas, Lithuania; dalius.jucius@ktu.lt (D.J.); valentinas.baltrusaitis@ktu.lt (V.B.); rimantas.gudaitis@ktu.lt (R.G.); igoris.prosycevas@ktu.lt (I.P.); brigita.abakeviciene@ktu.lt (B.A.); asta.guobiene@ktu.lt (A.G.); mindaugas.andrulevicius@ktu.lt (M.A.); viktoras.grigaliunas@ktu.lt (V.G.)
* Correspondence: algirdas.lazauskas@ktu.edu; Tel.: +370-671-73375

Received: 16 January 2019; Accepted: 1 February 2019; Published: 4 February 2019

Abstract: A photopolymerizable thiol-ene composition was prepared as a mixture of pentaerythritol tetrakis(3-mercaptopropionate) (PETMP) and 1,3,5-triallyl-1,3,5-triazine-2,4,6(1H,3H,5H)-trione (TTT), with 1 wt. % of 2,2-dimethoxy-2-phenylacetophenone (DMPA) photoinitiator. A systematic analytical analysis that investigated the crosslinked PETMP-TTT polymer coatings employed Fourier transform infrared spectroscopy, ultraviolet–visible spectroscopy, differential scanning calorimetry, thermogravimetric analysis, pencil hardness, thermo-mechanical cyclic tensile, scratch testing, and atomic force microscopy. These coatings exhibited high optical transparency and shape-memory that assisted scratch-healing properties. Scratches produced on the PETMP-TTT polymer coatings with different constant loadings (1.2 N, 1.5 N, and 2.7 N) were completely healed after the external stimulus was applied. The strain recovery ratio and total strain recovery ratio for PETMP-TTT polymer were found to be better than 94 ± 1% and 97 ± 1%, respectively. The crosslinked PETMP-TTT polymer network was also capable of initiating scratch recovery at ambient temperature conditions.

Keywords: photopolymerizable; thiol-ene network; scratch-healing; transparent

---

1. Introduction

Transparent polymer films are widely used as protective coatings in flat panel displays, touch screens, photovoltaic cells, and other devices. Accidental cuts and scratches tend to accumulate on the surface of transparent films and lead to the worsening of the optical transmission and distortion of displayed images. Thus, desirable but still challenging properties of such films are scratch resistance, with an ability to repair the damaged surface by self-healing.

Nowadays, there is a large variety of self-healing polymers, which can be classified into two broad categories: extrinsic and intrinsic self-healing materials [1,2]. Extrinsic self-healing polymers require inclusion of the specific healing agents that are loaded into microcapsules or vascular networks within a polymeric matrix. In this case, self-healing is triggered by the rupture of healant loaded vessels [3–5]. However, fabrication of the highly transparent extrinsic self-healing coatings are complicated, as various inclusions strongly scatter visible light and decrease optical transparency of the films [6]. On the contrary, intrinsic self-healing polymers are able to recover their properties due to the inherent physical interactions, such as molecular interdiffusion or reversible chemical bonds [4]. Reversible chemical bonding includes covalent bonds (i.e., dynamic bond exchange, Diels–Alder reactions, reversible C-ON bonds, photo-reversible reshuffling, and disulfide interchange), non-covalent interatomic bonds (metallic and ionic) and intermolecular forces (hydrogen and Van der

Waal's bonds) [4,7,8]. Exploration of reversible and adaptive noncovalent interactions has resulted in the development of highly complex chemical systems, e.g. supramolecular polymers, which have frequently been employed as self-healing materials [9,10]. Intrinsic self-healing polymers exhibit a latent self-healing functionality that is triggered by damage or by an outside stimulus (e.g. heat, light, or pressure) [11].

Recently, shape-memory polymers (SMPs) have attracted the attention of researchers, as materials capable of recovering their original shape after a temporary deformation when external stimulus is applied provides a mechanism to facilitate self-healing by bringing fractured surfaces into close proximity [12–14]. The main advantage of SMPs is the inherent shape recovery effect that eliminates the need for external force to partially or fully close the cracks, scratches, and other surface defects. However, it should be noted that in order to fully heal deep cuts the shape-memory effect is not sufficient and must be combined with other known intrinsic methods of self-healing [7].

A wide variety of polymers have been found to possess shape-memory properties. Among the them are amorphous covalently cross-linked (meth)acrylate-based shape-memory polymer networks prepared using free-radical polymerization, popular for their transparency and tunable properties [15,16]. Main drawbacks of the (meth)acrylate-based polymers include formation of heterogeneous polymer network and inhibition of the polymerization reaction by oxygen. Compared to commonly used (meth)acrylate-based SMPs, thiol-ene polymer systems present numerous advantages, including the negligible oxygen inhibition, low volume shrinkage, homogeneity of the polymer network, toughness, flexibility, and high optical transparency of the polymerized films [17–19].

The thiol-ene radical reaction (Scheme 1) is an organic reaction that involves the addition of a thiol to an alkene molecule to form an alkyl sulfide, also referred to as hydrothiolation [20]:

$$R^1-SH + \overset{}{\diagup}R^2 \longrightarrow R^1\diagdown_S\diagup R^2$$

Scheme 1. The thiol-ene radical reaction, produced from [20], with permission from © The Royal Society of Chemistry.

It commonly proceeds through the photochemical radical-initiated step-growth mechanism (Scheme 2), where thiyl radical from thiol component adds across a vinyl functional group of alkene, followed by the hydrogen abstraction from a thiol functional group resulting in the formation of carbon-centered radical, which undergoes chain transfer to a thiol group, regenerating the thiol radical [20,21]. This reaction cycle continues until one component is completely consumed.

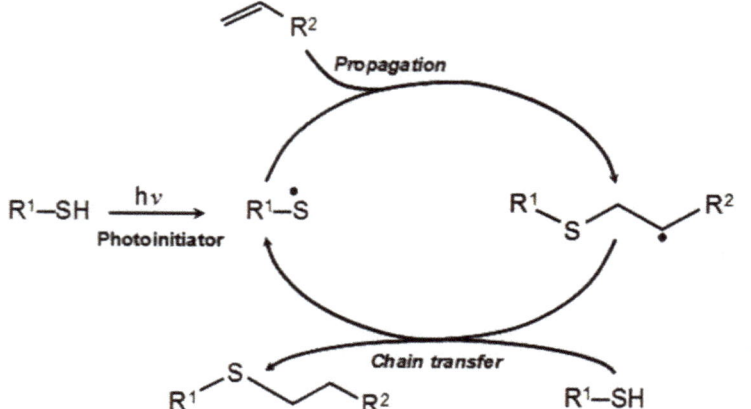

Scheme 2. The photochemical radical-initiated step-growth mechanism of the thiol-ene polymerization reaction, produced from [20], with permission from © The Royal Society of Chemistry.

Thiol-ene reactions are very fast and can complete in a matter of seconds. Furhter, they can withstand mild reaction conditions, as well as react with a little photoinitiator and can proceed without solvent that can be easily removed. Moreover, they do not require expensive transition-metals as catalysts [22–24]. The reaction proceeds more efficiently at pH in the range of 4–7 [25]. Thiol-ene reaction efficiency and kinetics are also highly dependent on the structure of the alkene moiety, where the reactivity is the greatest with strained and electron rich alkenes [26–28]. It is one of the most widely used reactions to prepare new crosslinked (e.g. tri- or tetrafunctional thiols and enes) or linear (using dithiols and dienes) branched polymeric structures with relatively narrow glass transition temperature range [29–32]. A common photoinitiator used in the thiol-ene reactions is 2,2-dimethoxy-2-phenylacetophenone. It gives a benzoyl radical and a tertiary carbon-centered radical which can insert directly into a carbon–carbon alkene bond or abstract a hydrogen atom from a thiol group carbon radical, which starts characteristic thiol-ene free-radical chain reaction [20].

Herein, we designed the thiol-ene shape-memory assisted scratch-healing polymer system that responds not only to the temperature changes as the external stimulus but is also capable of initiating a scratch recovery at ambient temperature conditions. Importantly, the developed thiol-ene polymer network exhibits high optical transmittance and holds a great potential as a high-performance flexible transparent material for optoelectronic applications.

2. Materials and Methods

2.1. Materials

All reagents and solvents were obtained at the highest purity and used without further purification unless otherwise specified. For instance, 1,3,5-triallyl-1,3,5-triazine-2,4,6(1H,3H,5H)-trione (TTT, trifunctional allyl component), pentaerythritol tetrakis(3-mercaptopropionate) (PETMP, tetrafunctional thiol component), and 2,2-dimethoxy-2-phenylacetophenone (DMPA, photoinitiator) were obtained from Sigma-Aldrich. Chemical structures of the trifunctional allyl, tetrafunctional thiol components, and photoinitiator are shown in Scheme 3.

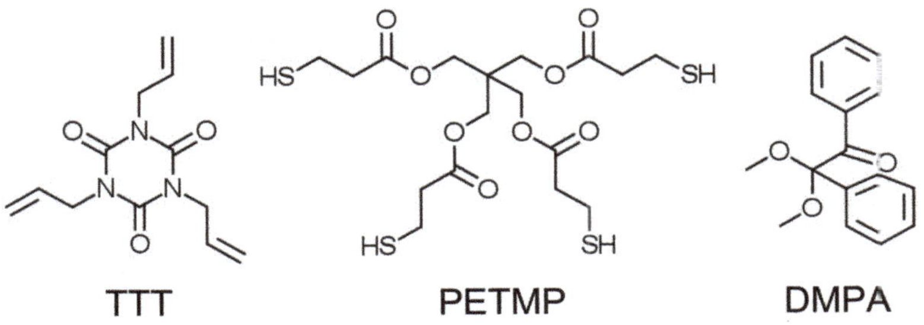

Scheme 3. Chemical structures of the starting compounds. These structures can be also found via Sigma-Aldrich search engine. (TTT, trifunctional allyl component); (PETMP, tetrafunctional thiol component); and (DMPA, photoinitiator).

2.2. Preparation of the Thiol-Ene PETMP-TTT Networks

Photopolymerizable thiol-ene composition was prepared as a mixture of PETMP and TTT with 1:1 stoichiometric ratio of thiol to ene functional groups, containing 1 wt. % of DMPA. The reason we chose cleavage photoinitiator is that it gives higher quantum yield for the production of reactive radicals as compared to the hydrogen-transfer photoinitiators [20]. Photoinitiator was dissolved in a warm PETMP at 60 °C in an amber glass jar, then the calculated amount of TTT was added avoiding the direct day or artificial light; components were thoroughly mixed with a spatula. The clear colorless viscous mixtures were applied on flexible polyethylene terephthalate (PET) substrates (APLI paper S.A., product Ref. 10580) as a 100 μm thick layer via the Meyer rod coating method. The PET substrate (thickness 0.1 mm) side was suitable for inkjet printing and was used in the deposition process. The water contact angle (CA) for this substrate side was determined to be 30 ± 1°, which is significantly lower than for polydopamine-coated (CA = 49.8°) or carboxyl-group-modified (CA = 50.6°) PET [33], but close to the O_2 plasma treated PET films (CA = 34 ± 1°) [34]. In another instance, deposition was performed on the polytetrafluoroethylene (PTFE) plate. Samples were cured simultaneously at the intensity of 1.64 mW/cm^2 (wavelength: 254 nm) and 0.8 mW/cm^2 (wavelength: 365 nm). After that, cured network was obtained, denoted as PETMP-TTT. Free-standing PETMP-TTT films were obtained by gently peeling the film from the PTFE plate.

2.3. Characterization

A Vertex 70 Fourier transform infrared (FTIR) spectrometer (Bruker Optics Inc., Ettlingen, Germany) equipped with a 30Spec (Pike Technologies) specular reflectance accessory having a fixed 30° angle of incidence (3/16" sampling area mask), was used to record the spectra. The sample was laid face down across the top of the 30Spec accessory and the spectrum of the sample was recorded at a resolution of 4 cm^{-1}. The software OPUS 6.0 (Bruker Optics Inc.) was used for data processing of the baseline correction of spectra.

Differential scanning calorimetry (DSC) measurements were carried out by using a Q2000 thermosystem (TA Instruments). The samples were examined at a heating/cooling rate of 10 °C/min under nitrogen atmosphere. Thermogravimetric analysis (TGA) was performed on a Q50 analyzer (TA Instruments). The heating rate was 10 °C/min under nitrogen atmosphere.

The pencil hardness test was used to measure the hardness of the PETMP-TTT coating, according to the standard ASTMD 3363. A vertical load of 750 g was applied at an angle of 45° to the horizontal coating surface, as the pencil was moved over the sample. The grade of coating was judged by the worn surfaces immediately after the pencil hardness tests.

Optical properties of the PETMP-TTT coatings were evaluated by measuring ultraviolet–visible (UV-Vis) transmission. Measurements were conducted using a fiber optic UV/VIS/NIR Spectrometer AvaSpec-2048 (Avantes, Apeldoorn, the Netherlands) in the wavelength range from 300 to 800 nm, with a resolution of 1.4 nm.

Thermo-mechanical cyclic tensile testing was performed using machine H10KT (Tinius Olsen, Kongsberg, Norway) equipped with a temperature controllable chamber. The sample was first heated to 70 °C for 120 s and then strained to 1.0% at a speed of 20 mm/min. After that, the sample was cooled to room temperature while 1.0% of the strain was kept for 10 min to fix temporary elongation. Next, the lower clamp returned to the original position. Once the force on the sample was released, it was heated again to 70 °C in order to recover; that is when the second cycle started. This cycle was repeated three times and the stress-strain curves were recorded for analysis. The maximum strain in the cyclic tensile tests and the residual strain after recovering in the Nth cycle were determined from stress-strain curves and used to calculate the strain recovery ratio R_r, as well as the total strain recovery ratio $R_{r,tot}$ after N passed cycles [35]. For the equations of R_r and $R_{r,tot}$ (i.e., Equations (1) and (2)), please refer to [35].

Scratch testing of PETMP-TTT coatings was performed with a custom-made PC controlled scratch testing apparatus. During the scratch test the PETMP-TTT coatings were scratched (scratch length 10 mm and speed 0.2 mm/s) with a sphero-conical stylus (cone angle 90° and indenter radius 45 μm) applying the constant loading of 1.2 N, 1.5 N, and 2.7 N, respectively. The scratches were performed in air atmosphere (temperature 23 °C and humidity 40%). The B-600MET series upright metallurgical microscope (OPTIKA Srl, Ponteranica, Italy) with a c-mount 2560 × 1920 resolution (5.0 Mpixel) camera (Optikam Pro 5LT) was used for the inspection of a scratch track before and after thermal treatment (70 °C for 5 min) of the coating. Inspection was performed immediately after the scratch test. The optical images of the scratch tracks were converted to greyscale with brightness and contrast levels equalized for each image, respectively. Scratch track profiles were obtained using a precision surface roughness tester TR200 (SaluTron Messtechnik GmbH, Frechen, Germany). In another instance, sphero-conical stylus was replaced with 19 mm precision stainless steel miniature wire cup brush with 1/8 inch shank and the surface of the coating was brushed applying the constant loading of 1.2 N and speed of 0.2 mm/s for two cycles (i.e. forward and backwards = 1 cycle). Brushing track length was 10 mm. Afterwards, time-lapse optical microscopy inspection was performed in order to reveal the self-healing properties of PETMP-TTT coatings; no external stimulus for the coating was applied in this case.

Surface morphology of PETMP-TTT coatings was investigated using atomic force microscopy (AFM). AFM experiments were carried out at room temperature using a NanoWizardIII atomic force microscope (JPK Instruments, Bruker Nano GmbH, Berlin, Germany), while the data was analyzed using a SurfaceXplorer and JPKSPM Data Processing software (Version spm-4.3.13, JPK Instruments, Bruker Nano GmbH). The AFM images were collected using a V-shaped silicon cantilever (spring constant of 3 N/m, tip curvature radius of 10.0 nm and the cone angle of 20°) operating in AC mode.

3. Results and Discussion

First, FTIR spectroscopy was used to determine optimal UV curing time for PETMP-TTT. Figure 1 shows the FTIR spectra of PETMP-TTT reaction mixture for the different UV curing time. The absorption peak of the S–H stretching band v_{SH} is located at 2570 cm^{-1} [36]. After 30 s, there was a significant decrease of v_{SH} (Figure 1b), indicating that thiol consumption proceeds very quickly. With the progress of the reaction, v_{SH} slightly decreases until UV-curing time is at 120 s. After that, no change of the v_{SH} intensity was observed, suggesting that the thiol conversion was complete. Accordingly, UV curing time of 120 s was chosen to form PETMP-TTT networks.

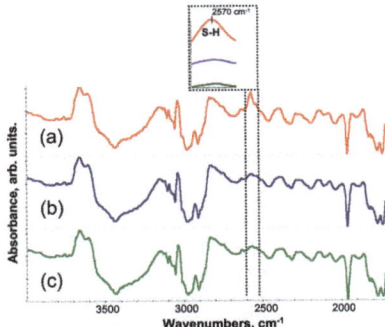

Figure 1. Fourier transform infrared (FTIR) spectra of PETMP-TTT reaction mixture for different UV curing time: (**a**) 0 s, (**b**) 30 s, and (**c**) 120 s. Inset: evolution of S–H band.

Figure 2a shows the DSC curves of the crosslinked PETMP-TTT polymer network. The high thiol functionality and high conversion resulted in relatively high glass transition temperature (T_g), with a value of 41 ± 1 °C and supported the FTIR results. This result is in good agreement with the T_g value previously reported in [37]. Thermal stability of the crosslinked PETMP-TTT polymer network was investigated by evaluating the weight loss behavior via TGA under nitrogen atmosphere. The quantitative properties such as the resultant weight loss temperature ($T_{d5\%}$) and the temperature corresponding to the maximum weight loss (T_{dmax}) were determined from TGA curve and are shown in Figure 2b. It can be seen that PETMP-TTT polymer network exhibits high thermal stability with $T_{d5\%}$ = 364 °C, which can be attributed to high thiol functionality resulting in high crosslinking density in the polymer network, which is in agreement with the FTIR results. The high T_{dmax} (444 °C) shows that the crosslinked polymer network has a very good heat resistance. It may originate from the nitrogen heterocycle in PETMP-TTT system.

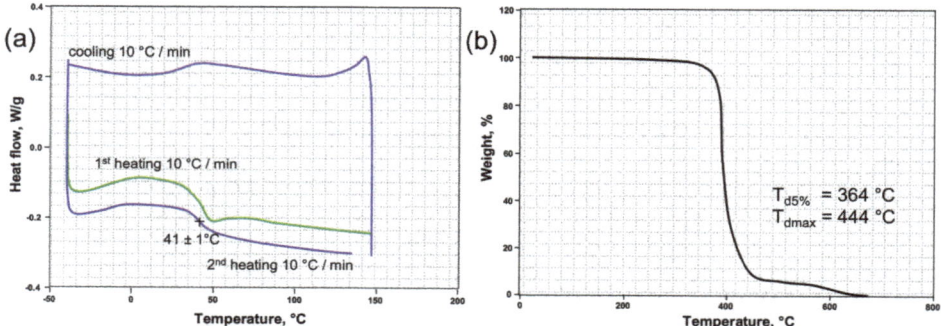

Figure 2. Differential scanning calorimetry (DSC) (**a**) and thermogravimetric analysis (TGA) (**b**) curves of the crosslinked PETMP-TTT polymer network.

Fabricated PETMP-TTT coatings could be characterized as glossy transparent films that ensure a smooth, comfortable touch feeling. The surface hardness of cured PETMP-TTT coatings was measured using pencil hardness test. Figure 3 shows the characteristic optical microscope digital photographs of PETMP-TTT coatings after the pencil hardness test. It was found that the pencil hardness value of 3B left no scratches on the coating surface—successfully passing the test—while the pencil hardness value of 2B failed. Shin, Junghwan, et al., who fabricated thiol-isocyanate-ene ternary networks by the sequential dual cure system, have also reported on similar pencil hardness value for polymer network with SH/C=C/NCO functional group molar ratio of 100:100:0 [38].

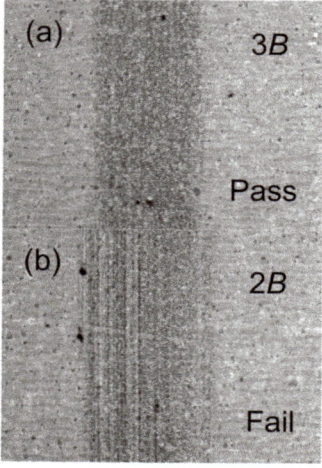

Figure 3. Characteristic optical microscope digital photographs (magnification 150×) of PETMP-TTT coatings after pencil hardness test: (**a**) 3B passed test and (**b**) 2B failed test.

From the digital photo (Figure 4), it can be directly seen that the crosslinked PETMP-TTT polymer network has very good transparency. Specifically, PETMP-TTT coatings show high transmittance in the wavelength range of 400–800 nm. At 300 nm, the transmittance of PETMP-TTT polymer network on PET substrate is close to zero, indicating very good blocking effect on medium and short-wave UV light (UVB and UVC bands).

In order to verify shape-memory effect in the fabricated PETMP-TTT polymer film, a simple test was performed. Free-standing PETMP-TTT film was folded to a certain shape and while this shape was fixed, the PETMP-TTT polymer was cooled down below 0 °C and left to stand at room temperature for 5 min. Afterwards, a free-standing PETMP-TTT film was heated with hot air (50–60 °C) and the shape recovery process was recorded. Figure 5 shows the shape recovery process of the free-standing PETMP-TTT film at different time instances and when heated with a hot air. Before the heating started, the vitrified film "remembered" its temporary shape (Figure 5a) so that only minor instantaneous recovery could occur at an ambient temperature. Heating above T_g triggered a fast return from temporary to the permanent shape through mobilization of the constituent polymer chains and release of latent strain energy. Full shape recovery of the tested film was achieved within 10 to 12 s. This result clearly indicated that crosslinked PETMP-TTT polymer network possesses shape-memory properties.

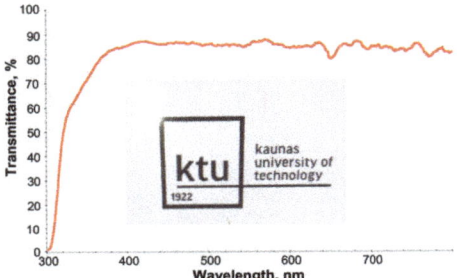

Figure 4. UV-vis spectrum and digital photo of PETMP-TTT coating on PET substrate over KTU logo, displayed through a computer monitor.

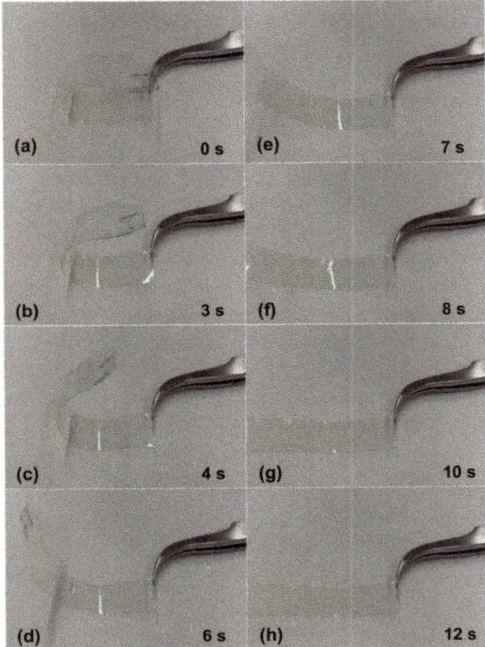

Figure 5. Shape-memory test of the crosslinked PETMP-TTT polymer free-standing film: (**a**) Digital photograph of free-standing PETMP-TTT film with fixed shape after cooling down below 0 °C and (**b**), (**c**), (**d**), (**e**), (**f**), (**g**), and (**h**) shape recovery process at different time instances when heated with hot air.

For further quantification of shape-memory's effect on the thermo-mechanical cyclic, tensile testing was performed. Figure 6 shows shape-memory behavior of the free-standing PETMP-TTT film, with characteristic cyclic tensile stress-strain curves. The R_r and $R_{r,tot}$ values for PETMP-TTT polymer were found to be better than 94 ± 1% and 97 ± 1%, respectively, which indicates very good shape-memory behavior comparable with recently published results for PEG blends [39]. A slight improvement of R_r was observed with increase in the number of applied cycles: R_r increased from 94 ± 1% (first cycle) to 96 ±1% after three cycles of the tensile test. It can be considered as a normal process for cyclic thermo-mechanical investigations, where the first few cycles often tend to differ from each other because of the history of the film and reorganization of the polymer chains [40].

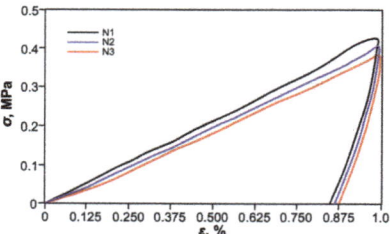

Figure 6. Cyclic tensile stress-strain curves of the crosslinked PETMP-TTT polymer network.

Scratch testing was performed in order to investigate the scratch-healing properties of PETMP-TTT coatings after external stimulus was applied. Figure 7 shows optical microscope digital photographs of scratch tracks obtained on the coating with different constant loading as well as corresponding healing process of the scratch after 5 min at 70 °C. A scratch track profile area was used to calculate a scratch-healing ratio (S_r) of PETMP-TTT coatings in Equation (1)

$$S_r = \frac{A_1 - A_2}{A_1} \times 100\% \tag{1}$$

where A_1 stands for the scratch track profile area after scratch, while A_2 stands for the scratch track profile area after healing. Indicators for the beginning and the end of scratch are added along the profiles. Scratch tracks obtained with the constant loading of 1.2 N and 1.5 N were almost completely healed after the thermal treatment of the PETMP-TTT coating with an S_r of 96 and 94%. It is important to note that no decomposition of polymer was observed during the thermal treatment process. Furthermore, the crosslinked PETMP-TTT network also exhibited considerable scratch-healing (S_r = 91%) properties of the scratches obtained with a higher constant loading of 2.7 N, as observed in Figure 7c$_1$ and c$_2$.

In another instance, PETMP-TTT coatings were cooled down below 0 °C and left to stand at a room temperature for 5 min prior to the scratch testing. This procedure was performed in order to immobilize the polymer chains in covalently bonded 3D network and ensure improved form fixity. The testing procedure was analogous to the previously described scratch test. Figure 8 shows optical microscope digital photographs of scratch tracks obtained on the quenched coatings with different constant loading as well as corresponding healing process of the scratches after 5 min at 70 °C. It is evident that cooling down prior to the scratch testing resulted in a more efficient shape recovery of the scratched films. In all cases, the scratches were completely healed with a S_r value of 99%. It is suggested that a quenching of the polymerized PETMP-TTT network serves as a freezing process for the internal strain energy, which is the primary driving force for the shape recovery of the scratched polymeric films.

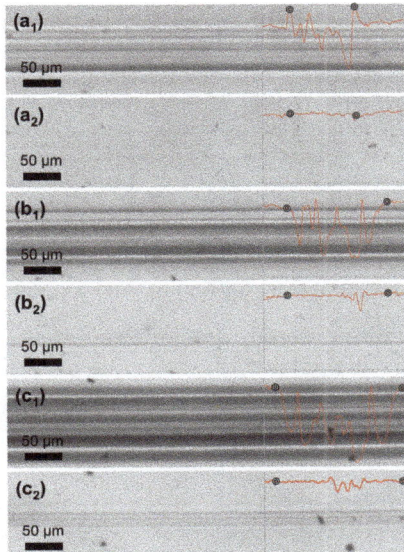

Figure 7. Scratch-healing of the crosslinked PETMP-TTT polymer coatings: (a_1), (b_1), and (c_1) optical microscope digital photographs of characteristic scratch track sections before, and (a_2), (b_2), and (c_2) after the healing process of the scratch with different constant loading of 1.2 N, 1.5 N, and 2.7 N, respectively. Insets show characteristic scratch track profiles with corresponding indicators for the beginning and the end of scratch.

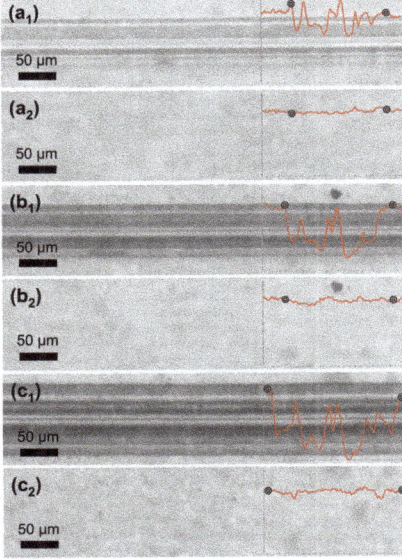

Figure 8. Shape-memory assisted scratch-healing of the quenched PETMP-TTT polymer coatings: (a_1), (b_1), and (c_1) optical microscope digital photographs of characteristic scratch track sections before, and (a_2), (b_2), and (c_2) after the healing process of the scratch with different constant loading of 1.2 N, 1.5 N, and 2.7 N, respectively. Insets show characteristic scratch track profiles with corresponding indicators for the beginning and the end of scratch.

Further, the scratch self-repairing ability of PETMP-TTT coatings was investigated at room temperature and without external stimulus applied. In this case, the surface brushing tests with a stainless-steel miniature wire brush were performed. Figure 9 presents evolution of the initial film scratches with a time at ambient temperature. It can be seen that noticeable scratches were made by the brush wires having lateral dimensions of the order of tens of micrometers almost completely disappear within 110 min time span even at a room temperature. Although the PETMP-TTT coatings, that are heated above the T_g, initiate the shape-memory actuation response more rapidly, the healing process for the brushed specimens is also quite effective. It can be associated with delayed elasticity of the distorted polymer network. When PETMP-TTT polymer is heated above the T_g, the segmental mobility of the network is higher [41], resulting in a faster self-healing process.

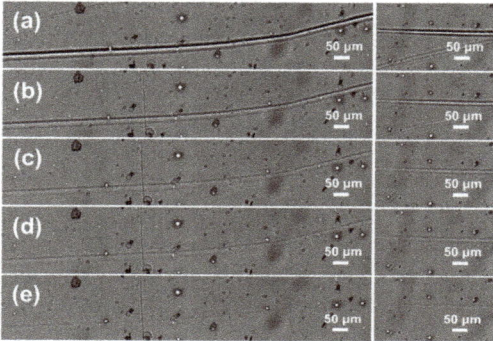

Figure 9. Scratch-healing of the crosslinked PETMP-TTT polymer network: (**a**) initial scratches obtained after brushing; (**b**), (**c**), (**d**), and (**e**) images of the scratches after 10 min, 30 min, 50 min, and 110 min at room temperature, respectively.

AFM measurements were performed to investigate the surface morphology of PETMP-TTT polymer coating at the nanoscale (Figure 10). The topography of PETMP-TTT shows a homogeneous surface with 22.0 ± 2° (i.e., the most frequent orientation) oriented morphological features having a mean height of 3.64 nm and the root mean square roughness (R_q) of 0.40 nm. The surface of PETMP-TTT is dominated by the valleys with skeweness (R_{sk}) value of -0.93 and has a leptokurtoic distribution of the morphological features with kurtosis (R_{ku}) value of 5.76, indicating many high peaks and low valleys.

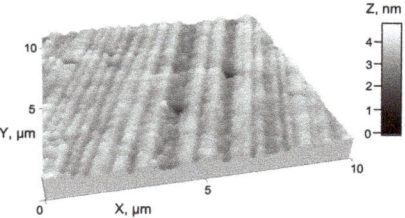

Figure 10. Characteristic atomic force microscopy (AFM) topographical image with normalized Z axis in nm of PETMP-TTT polymer coating.

With the emergence of new optoelectronic products like foldable smartphones and wearable electronic sensors with the flexible electronic components, scratch-healing properties of the materials acting as protective coatings will be more and more desirable. The particular crosslinked PETMP-TTT polymer network (Figure 11) that exhibits high optical transparency and efficient shape-memory assisted scratch-healing properties could be considered as a functional layer for the structured optoelectronic devices. Importantly, this is a significant finding, as the scratch-healing properties for PETMP-TTT polymer network were not explored previously. Synthesis, deposition, and curing technological procedures of PETMP-TTT coatings are not sophisticated, time efficient, and scalable, and thus could be easily integrated in the large area flexible electronic manufacturing processes.

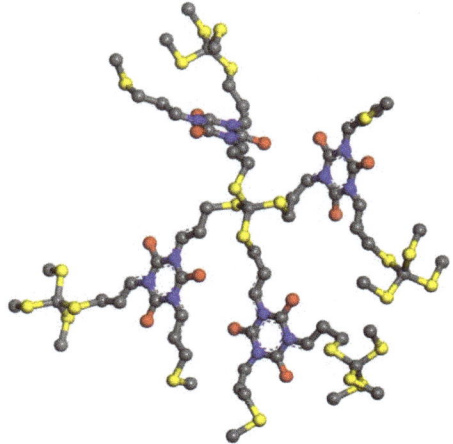

Figure 11. Representation of the crosslinked PETMP-TTT polymer network fragment.

4. Conclusions

The PETMP and TTT, acting as tetrafunctional thiol and trifunctional allyl components, respectively, were used in the preparation of photopolymerizable thiol-ene composition, with 1 wt. % of DMPA photoinitiator. An optimal UV-curing time of 120 s was determined using FTIR spectra analysis, and in particular, monitoring the intensity change of an S–H stretching band at 2570 cm^{-1}. The crosslinked PETMP-TTT polymer network exhibited a relatively high glass transition temperature (T_g) with a value of 41 ± 1 °C, indicating high thiol functionality and high conversion. TGA measurements showed that PETMP-TTT polymer network exhibits a high thermal stability, with $T_{d5\%}$ = 364 °C, which is a result of the high crosslinking density in the polymer network. The PETMP-TTT coatings passed the 3B pencil hardness test. The shape-memory test revealed that PETMP-TTT polymer network possesses shape-memory properties. Further quantification of the shape-memory behavior of the crosslinked PETMP-TTT polymer was obtained using cyclic tensile testing. The determined R_r and $R_{r,tot}$ values were found to be better than 94 ± 1% and 97 ± 1%, respectively. It was found that scratches produced with a different constant loading of 1.2 N, 1.5 N, and 2.7 N on the PETMP-TTT coating heal more efficiently with a quenching procedure applied prior to the scratch testing. In this case, the scratches were completely healed with S_r value of 99% for the all scratch testing instances. It was also found that the crosslinked PETMP-TTT polymer network was also capable to initiate a scratch recovery at ambient temperature conditions, presumably due to the delayed elasticity of the distorted polymer network. It was suggested that a crosslinked PETMP-TTT polymer network could be integrated as a scratch-healing and transparent coating in a new generation of the optoelectronic products.

Author Contributions: Conceptualization, A.L., V.B., and V.G.; Methodology, A.L., V.B., and I.P.; Validation, A.L. and D.J.; Formal Analysis, A.L. and D.J.; Investigation, A.L., R.G., I.P., B.A., A.G., and M.A.; Resources, A.L.; Writing—Original Draft Preparation, A.L. and D.J.; Writing—Review and Editing, A.L., V.G. and D.J.; Visualization, A.L.; Supervision, A.L.; Project Administration, A.L. and B.A.; Funding Acquisition, A.L.

Funding: This research was (and is) funded by the European Social Fund under the No 09.3.3-LMT-K-712-01 "Improvement of researchers' qualification by implementing world-class R&D projects" measure. Grant No. 09.3.3-LMT-K-712-01-0074.

Acknowledgments: Special thanks goes to Jūratė Simokaitienė from Kaunas University of Technology.

Conflicts of Interest: The authors declare no conflict of interest.

References

1. Yuan, Y.; Yin, T.; Rong, M.; Zhang, M. Self healing in polymers and polymer composites. Concepts, realization and outlook: A review. *Express Polym. Lett.* **2008**, *2*, 238–250. [CrossRef]
2. Bekas, D.; Tsirka, K.; Baltzis, D.; Paipetis, A. Self-healing materials: A review of advances in materials, evaluation, characterization and monitoring techniques. *Compos. Part B-Eng.* **2016**, *87*, 92–119. [CrossRef]
3. Lee, M.W.; An, S.; Yoon, S.S.; Yarin, A.L. Advances in self-healing materials based on vascular networks with mechanical self-repair characteristics. *Adv. Colloid Interface Sci.* **2017**, *252*, 21–37. [CrossRef]
4. Thakur, V.K.; Kessler, M.R. Self-healing polymer nanocomposite materials: A review. *Polymer* **2015**, *69*, 369–383. [CrossRef]
5. An, S.; Lee, M.W.; Yarin, A.L.; Yoon, S.S. A review on corrosion-protective extrinsic self-healing: Comparison of microcapsule-based systems and those based on core-shell vascular networks. *Chem. Eng. J. (Lausanne)* **2018**, *344*, 206–220. [CrossRef]
6. Amendola, V.; Meneghetti, M. Advances in self-healing optical materials. *J. Mater. Chem.* **2012**, *22*, 24501–24508. [CrossRef]
7. Zhang, P.; Li, G. Advances in healing-on-demand polymers and polymer composites. *Prog. Polym. Sci.* **2016**, *57*, 32–63. [CrossRef]
8. Yang, Y.; Urban, M.W. Self-healing polymeric materials. *Chem. Soc. Rev.* **2013**, *42*, 7446–7467. [CrossRef]
9. Van Gemert, G.M.; Peeters, J.W.; Söntjens, S.H.; Janssen, H.M.; Bosman, A.W. Self-healing supramolecular polymers in action. *Macromol. Chem. Phys.* **2012**, *213*, 234–242. [CrossRef]
10. Liu, K.; Kang, Y.; Wang, Z.; Zhang, X. 25th anniversary article: Reversible and adaptive functional supramolecular materials: "Noncovalent interaction" matters. *Adv. Mater.* **2013**, *25*, 5530–5548. [CrossRef]
11. Geitner, R.; Legesse, F.B.; Kuhl, N.; Bocklitz, T.W.; Zechel, S.; Vitz, J.; Hager, M.; Schubert, U.S.; Dietzek, B.; Schmitt, M. Do You Get What You See? Understanding Molecular Self-Healing. *Chem.–Eur. J.* **2018**, *24*, 2493–2502. [CrossRef] [PubMed]
12. Hager, M.D.; Bode, S.; Weber, C.; Schubert, U.S. Shape memory polymers: Past, present and future developments. *Prog. Polym. Sci.* **2015**, *49*, 3–33. [CrossRef]
13. Luo, X.; Mather, P.T. Shape memory assisted self-healing coating. *ACS Macro Lett.* **2013**, *2*, 152–156. [CrossRef]
14. Wypych, G. *Self-Healing Materials: Principles and Technology*; Elsevier: Amsterdam, The Netherlands, 2017.
15. Ortega, A.M.; Yakacki, C.M.; Dixon, S.A.; Likos, R.; Greenberg, A.R.; Gall, K. Effect of crosslinking and long-term storage on the shape-memory behavior of (meth) acrylate-based shape-memory polymers. *Soft Matter* **2012**, *8*, 7381–7392. [CrossRef]
16. Song, L.; Hu, W.; Wang, G.; Niu, G.; Zhang, H.; Cao, H.; Wang, K.; Yang, H.; Zhu, S. Tailored (Meth) Acrylate Shape-Memory Polymer Networks for Ophthalmic Applications. *Macromol. Biosci.* **2010**, *10*, 1194–1202. [CrossRef] [PubMed]
17. Lowe, A.B. Thiol-ene "click" reactions and recent applications in polymer and materials synthesis. *Polym. Chem.* **2010**, *1*, 17–36. [CrossRef]
18. Schreck, K.M.; Leung, D.; Bowman, C.N. Hybrid organic/inorganic thiol–ene-based photopolymerized networks. *Macromolecules* **2011**, *44*, 7520–7529. [CrossRef] [PubMed]
19. Nair, D.P.; Cramer, N.B.; Scott, T.F.; Bowman, C.N.; Shandas, R. Photopolymerized thiol-ene systems as shape memory polymers. *Polymer* **2010**, *51*, 4383–4389. [CrossRef]
20. Bordoni, A.V.; Lombardo, M.V.; Wolosiuk, A. Photochemical radical thiol–ene click-based methodologies for silica and transition metal oxides materials chemical modification: A mini-review. *RSC Adv.* **2016**, *6*, 77410–77426. [CrossRef]
21. Kloxin, C.J.; Scott, T.F.; Bowman, C.N. Stress relaxation via addition− fragmentation chain transfer in a thiol-ene photopolymerization. *Macromolecules* **2009**, *42*, 2551–2556. [CrossRef]
22. Montanez, M.I.; Campos, L.M.; Antoni, P.; Hed, Y.; Walter, M.V.; Krull, B.T.; Khan, A.; Hult, A.; Hawker, C.J.; Malkoch, M. Accelerated growth of dendrimers via thiol− ene and esterification reactions. *Macromolecules* **2010**, *43*, 6004–6013. [CrossRef]

23. Barner-Kowollik, C.; Du Prez, F.E.; Espeel, P.; Hawker, C.J.; Junkers, T.; Schlaad, H.; Van Camp, W. "Clicking" polymers or just efficient linking: What is the difference? *Angew. Chem. Int. Ed.* **2011**, *50*, 60–62. [CrossRef] [PubMed]
24. Zuo, Y.; Lu, H.; Xue, L.; Wang, X.; Ning, L.; Feng, S. Preparation and characterization of luminescent silicone elastomer by thiol–ene "click" chemistry. *J. Mater. Chem. C* **2014**, *2*, 2724–2734. [CrossRef]
25. Colak, B.; Da Silva, J.C.; Soares, T.A.; Gautrot, J.E. Impact of the molecular environment on thiol–ene coupling for biofunctionalization and conjugation. *Bioconjugate Chem.* **2016**, *27*, 2111–2123. [CrossRef] [PubMed]
26. Fisher, S.A.; Baker, A.E.; Shoichet, M.S. Designing peptide and protein modified hydrogels: Selecting the optimal conjugation strategy. *J. Am. Chem. Soc.* **2017**, *139*, 7416–7427. [CrossRef] [PubMed]
27. Hoyle, C.E.; Bowman, C.N. Thiol–ene click chemistry. *Angew. Chem. Int. Ed.* **2010**, *49*, 1540–1573. [CrossRef] [PubMed]
28. Cramer, N.B.; Reddy, S.K.; O'Brien, A.K.; Bowman, C.N. Thiol– ene photopolymerization mechanism and rate limiting step changes for various vinyl functional group chemistries. *Macromolecules* **2003**, *36*, 7964–7969. [CrossRef]
29. Khire, V.S.; Lee, T.Y.; Bowman, C.N. Synthesis, characterization and cleavage of surface-bound linear polymers formed using thiol– ene photopolymerizations. *Macromolecules* **2008**, *41*, 7440–7447. [CrossRef]
30. Li, Y.-h.; Wang, D.; Buriak, J.M. Molecular Layer Deposition of Thiol– Ene Multilayers on Semiconductor Surfaces. *Langmuir* **2009**, *26*, 1232–1238. [CrossRef]
31. Hoyle, C.E.; Lee, T.Y.; Roper, T. Thiol–enes: Chemistry of the past with promise for the future. *J. Polym. Sci. Part A Pol. Chem.* **2004**, *42*, 5301–5338. [CrossRef]
32. Durham, O.Z.; Norton, H.R.; Shipp, D.A. Functional polymer particles via thiol–ene and thiol–yne suspension "click" polymerization. *RSC Advances* **2015**, *5*, 66757–66766. [CrossRef]
33. Zhao, C.; Xing, L.; Xiang, J.; Cui, L.; Jiao, J.; Sai, H.; Li, Z.; Li, F. Formation of uniform reduced graphene oxide films on modified PET substrates using drop-casting method. *Particuology* **2014**, *17*, 66–73. [CrossRef]
34. Jucius, D.; Grigaliūnas, V.; Kopustinskas, V.; Lazauskas, A.; Guobienė, A. Wettability and optical properties of O2 and CF4 plasma treated biaxially oriented semicrystalline poly (ethylene terephthalate) films. *Appl. Surf. Sci.* **2012**, *263*, 722–729. [CrossRef]
35. Jankauskaitė, V.; Laukaitienė, A.; Mickus, K.V. Shape memory properties of poly (ε-caprolactone) based thermoplastic polyurethane secondary blends. *Strain* **2008**, *2*, 26.
36. Miao, J.-T.; Yuan, L.; Guan, Q.; Liang, G.; Gu, A. Water-Phase Synthesis of a Biobased Allyl Compound for Building UV-Curable Flexible Thiol-Ene Polymer Networks with High Mechanical Strength and Transparency. *ACS Sustain. Chem. Eng.* **2018**, *6*, 7902–7909. [CrossRef]
37. Chen, L.; Wu, Q.; Wei, G.; Liu, R.; Li, Z. Highly stable thiol–ene systems: From their structure–property relationship to DLP 3D printing. *J. Mater. Chem. C* **2018**, *6*, 11561–11568. [CrossRef]
38. Shin, J.; Matsushima, H.; Comer, C.M.; Bowman, C.N.; Hoyle, C.E. Thiol– isocyanate– ene ternary networks by sequential and simultaneous thiol click reactions. *Chem. Mater.* **2010**, *22*, 2616–2625. [CrossRef]
39. Boumezgane, O.; Messori, M. Poly (ethylene glycol)-based shape-memory polymers. *Int. J. Polym. Anal. Charact.* **2017**, *22*, 463–471. [CrossRef]
40. Lendlein, A.; Kelch, S. Shape-memory polymers. *Angew. Chem. Int. Ed.* **2002**, *41*, 2034–2057. [CrossRef]
41. Acosta Ortiz, R.; Acosta Berlanga, O.; García Valdez, A.E.; Aguirre Flores, R.; Télles Padilla, J.G.; Méndez Padilla, M.G. Self-healing photocurable epoxy/thiol-ene systems using an aromatic epoxy resin. *Adv. Mater. Sci. Eng.* **2016**, *2016*. [CrossRef]

© 2019 by the authors. Licensee MDPI, Basel, Switzerland. This article is an open access article distributed under the terms and conditions of the Creative Commons Attribution (CC BY) license (http://creativecommons.org/licenses/by/4.0/).

Article

Total Performance of Magneto-Optical Ceramics with a Bixbyite Structure

Akio Ikesue [1], Yan Lin Aung [1,*], Shinji Makikawa [2] and Akira Yahagi [2]

1. World-Lab. Co., Ltd., Mutsuno, Atsutaku, Nagoya 456-0023, Japan; poly-ikesue@s5.dion.ne.jp
2. Shin-Etsu Chemical Co., Ltd., Advanced Functional Materials Research Center, Matsuida, Annaka, Gunma 379-0224, Japan; s_makikawa@shinetsu.jp (S.M.); yahagi@shinetsu.jp (A.Y.)
* Correspondence: poly-yan@r2.dion.ne.jp

Received: 26 December 2018; Accepted: 25 January 2019; Published: 30 January 2019

Abstract: High-quality magneto-optical ceramics $(Tb_xY_{1-x})_2O_3$ ($x = 0.5$–1.0) with a Bixbyite structure were extensively investigated for the first time. The total performances of these ceramics were far superior to those of commercial TGG ($Tb_3Ga_5O_{12}$) crystal, which is regarded as the highest class of Faraday rotator material. In particular, the Verdet constant of Tb_2O_3 (when $x = 1.0$) ceramic was the largest—495 to 154 rad·T^{-1}·m^{-1} in the wavelength range of 633 to 1064 nm, respectively. It was possible to further minimize the Faraday isolator device. The insertion loss of this ceramic was equivalent to that of the commercial TGG single crystal (0.04 dB), and its extinction ratio reached more than 42 dB, which is higher than the value for TGG crystal (35 dB). The thermal lens effect (1/f) was as small as 0.40 m^{-1} as measured by a 50 W fiber laser. The laser damage threshold of this ceramic was 18 J/cm^2, which is 1.8 times larger than that of TGG, and it was not damaged during a power handling test using a pulsed laser (pulse width 50 ps, power density 78 MW/cm^2) irradiated at 2 MHz for 7000 h.

Keywords: faraday rotator material; optical isolator; transparent ceramics

1. Introduction

In 1995, highly efficient laser oscillation using a polycrystalline Nd:YAG ($Y_3Al_5O_{12}$) ceramic material was reported for the first time [1]. Since then, research and development on various types of ceramic laser materials and laser oscillation has been successively reported [2]. Generally, ceramic materials are easily influenced by Mie scattering and Rayleigh scattering [3–6] as they contain many grain boundaries which degrade the oscillation efficiency and laser beam quality when they are used as a laser gain medium. However, recent studies have revealed that certain types of ceramic materials can provide novel characteristics that cannot be achieved in single crystals [7–9]. Polycrystalline ceramics are anticipated to be widely applied to the field of photonics in addition to laser applications.

Faraday elements (Faraday rotator materials) that apply Faraday's effect are classified into either Fe-containing magnetic material [10,11] or paramagnetic material [12]. The former is represented by Bi-doped iron garnet and is commonly used in isolator applications owing to its high isolation performance. The latter—paramagnetic material—has been applied for wavelength regions lower than 1.2 µm where iron garnet cannot be used. Compared with those of the Fe-included magnetic type, paramagnetic type isolators have extremely small Verdet constants. Even the commercially widely used TGG ($Tb_3Ga_5O_{12}$) material with a Garnet structure has a low Verdet constant at only around 36–40 rad·T^{-1}·m^{-1} [13]. Consequently, the device needs to be of length 20 mm or more, and a larger magnetic field is necessary for isolator applications. The technical issues above still need to be solved. Although NTF ($Na_{0.37}Tb_{0.63}F_{2.26}$) for high-power applications [14,15] and another garnet material,

TSAG ($Tb_3Sc_2Al_5O_{12}$), have been reported [16], their Verdet constants are no more than 0.7 times and 1.3 times that of TGG, respectively.

In the case of the non-magnetic-type Faraday rotator, firstly, we considered that the content of Tb ions must be as high as possible in the materials and the crystal structure must be Bixbyite in order to improve the performance of the Faraday rotator materials, especially in terms of the Verdet constant. Additionally, we found out that the development of Tb_2O_3-Re_2O_3 (Re:Sc, Y, Lu, Gd, etc.) would be the most effective approach. However, since Tb_2O_3 is not stable at room temperature, when Tb_4O_7 is used as a starting material, (1) oxygen gas is released during the sintering process, i.e., $Tb_4O_7 \rightarrow 2Tb_2O_3 + 1/2O_2$, and the powder compact is finally decomposed into pieces; and (2) it appears to undergo phase transition from an orthorhombic $\Leftrightarrow$ cubic crystal system near 1400 °C to an orthorhombic $\Leftrightarrow$ hexagonal crystal system near 2100 °C [17,18]. These phase transitions cause volumetric changes in the material and induce mechanical stress, causing cracking, etc. Therefore, there is almost no possibility to produce this type of material using a conventional fabrication method such as melt growth or sintering technologies.

In this study, we investigated the technological difficulties in the fabrication of this material and systematically solved the above technical issues. High-quality transparent $(Tb_xY_{1-x})_2O_3$ (x = 0.5–1.0) sintered bodies (hereafter abbreviated as TYO) with a Bixbyite structure were successfully produced, and their total performances relating to their use in practical application as a Faraday rotator device were extensively characterized for the first time. These TYO ceramics have optical properties well suited for Faraday rotators, and due to their large Verdet constants, these new materials are very promising for the development of Faraday rotators with very compact and small size compared to the commercial TGG optical isolator.

2. Experimental Procedures

Tb_4O_7 (Shin-Etsu, Tokyo, Japan, RU, 99.99%) and Y_2O_3 (Shin-Etsu, RU, 99.999%) powders were used as starting materials. Tb_2O_3 (yellowish white color) powders were prepared by heat treating the dark brown Tb_4O_7 raw powders (Shin-Etsu, RU, 99.99%) under a hydrogen atmosphere. Ethanol (analytical reagent) was used as the solvent. A small amount of ZrO_2 (TOSOH, Tokyo, Japan, TZ-0, 99.9%: 0.5–1.5 mass %) was used as a sintering aid.

Tb_2O_3 powder was mixed with Y_2O_3 powder in ethanol solvent for 10 h by a conventional ball-milling process. The obtained slurry was dried and granulated by using a spray-dryer (SAKAMOTO Engineering, TRS-4W, Kawasaki, Japan). The premixed Tb_2O_3-Y_2O_3 powders were made into tablets using a metal mold (internal diameter: 8 mm) by uniaxial pressing (RIKEN, CDM-5PA, Tokyo, Japan). Then, the tablets were isostatically pressed in a CIP machine (cold isostatic press, KOBE Steel, P200, Tokyo, Japan) with a pressure of 196 MPa. Depending on the Tb doping level, these powder compacts were sintered in a vacuum furnace (W-heater, Special ordered furnace, Futek Furnace Inc., Yokohama, Japan) under a vacuum level of 1×10^{-3} Pa at 1500–1680 °C for 3 h. Then, the pre-sintered tablets were treated in an HIP (hot isostatic press, KOBE Steel, SYS50X-SB, Tokyo, Japan) machine with a temperature range from 1500 to 1700 °C for 2 h under Ar gas pressure at 176 MPa. The sintering temperature and pressure were adapted in accordance with the Tb content. Transparent ceramics were achieved after the HIP treatment. Their basic optical properties and magneto-optical properties were investigated. Details of the characterization are described in the Supplementary Materials. The technical issues relating to single-crystal TYO grown by the conventional melt growth method are also discussed in the Supplementary Materials.

To evaluate the Faraday rotation performance, the same experimental setup reported in a previous paper was used [19]. Samples of $(Tb_{0.6}Y_{0.4})_2O_3$ ceramics (5 mm in diameter by 8 mm length) and TGG single crystal (5 mm in diameter by 20 mm length, Electro-Optics Technology Inc., Traverse City, MI, USA) with <111> orientation were used. Each sample was clamped in a copper holder and put in a commercial Faraday rotator magnetic housing. The average magnetic field exerted on the TGG crystal and $(Tb_{0.6}Y_{0.4})_2O_3$ ceramics was 1 T. A polarization plane of laser light was rotated by the Faraday effect due to the magnetic field. The transmitted laser output was measured using a power meter.

3. Results

3.1. Synthesis and Characterization of the Novel Ceramic Faraday Rotator Material

A ceramic fabrication process was applied to produce sintered bodies with a $(Tb_{0.6}Y_{0.4})_2O_3$ composition. Tb_4O_7 (Shin-Etsu, RU, 99.99%) and Y_2O_3 (Shin-Etsu, RU, 99.999%) powders were used as starting materials; they were mixed in ethanol solvent for 10 h, then the dried premixed powders were pressed in a CIP (Cold Isostatic Press) machine at 196 MPa. The color of the obtained powder compacts was dark brown. Then, the powder compacts were sintered (1) under hydrogen atmosphere and (2) under vacuum (1×10^{-3} Pa) at 1600 °C for 2 h, separately. In both processes, almost all samples were crushed into pieces after sintering in hydrogen, and many cracks occurred after sintering in vacuum.

From the TG-DTA (Thermogravimetry Differential Thermal Analysis, Thermo plus EVO TG8120, RIGAKU, Akishima, Japan) analysis result, we concluded that the following oxygen dissociation reaction occurred during the heating of Tb_4O_7 due to the release of oxygen inside the material when the samples cracked or were crushed into pieces.

$$7Tb_4O_7 \Leftrightarrow 4Tb_7O_{12} + 1/2O_2 \text{ at } 540\ °C$$

$$2Tb_7O_{12} \Leftrightarrow 7Tb_2O_3 + 3/2O_2 \text{ at } 940\ °C$$

To avoid this cracking problem, first, Tb_2O_3 (yellowish white color) powders were prepared by heat treating the dark brown Tb_4O_7 raw powders (Shin-Etsu, RU, 99.99%) under a hydrogen atmosphere. The fabrication process for TYO ceramics derived from Tb_2O_3 is shown in Figure 1a. Then, the Tb_2O_3 powder was mixed with Y_2O_3 (Shin-Etsu, RU, 99.999%) powder in ethanol solvent for 10 h by a ball-milling process. The obtained slurry was dried and granulated using a spray-dryer. The premixed Tb_2O_3-Y_2O_3 powders were made into tablets using a metal mold (Φ 8 mm) by uniaxial pressing and a CIP (cold isostatic press) machine with a pressure of 196 MPa. The obtained powder compacts are shown in Figure 1b. Depending on the content of Tb, these power compacts were sintered under vacuum (1×10^{-3} Pa) conditions at 1500–1680 °C for 3 h. Then, the pre-sintered tablets were treated in an HIP (hot isostatic press) machine with a temperature range from 1500 to 1700 °C for 2 h under Ar gas pressure at 176 MPa. The sintering temperature and pressure were adapted in accordance with the Tb content. After the HIP treatment, yellowish transparent ceramics were achieved (see Figure 1c). The color of the transparent ceramic samples varied with the content of Tb ions. The higher the Tb content was, the deeper the color of the sample. However, when pure Tb_2O_3 was produced, it was colorless and transparent. If there were no sintering aids used, the grain growth was accelerated during sintering, and the sintered sample showed poor translucency or cracks occurred inside the samples, or it ended up cracking into pieces in the worst case (see Figure 1d).

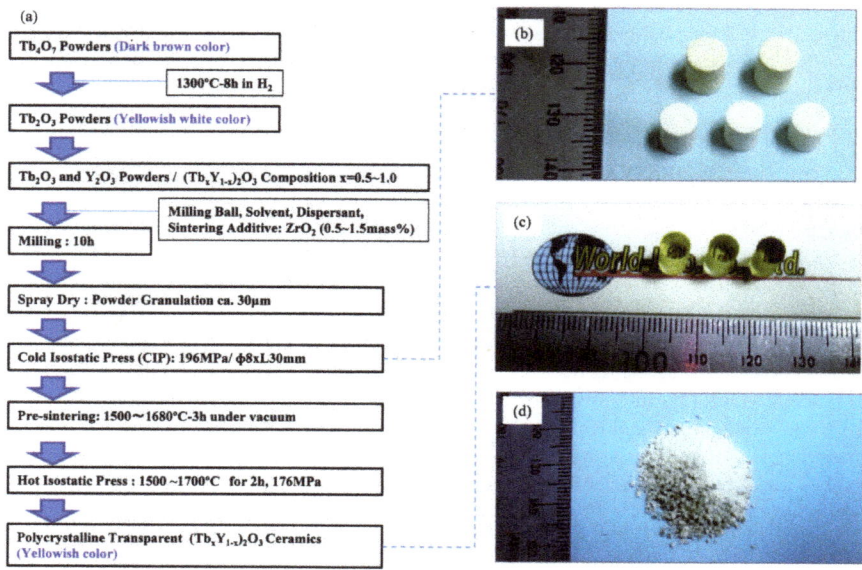

Figure 1. (**a**) Fabrication process for $(Tb_xY_{1-x})_2O_3$ ceramics derived from Tb_2O_3 and Y_2O_3 raw powders; (**b**) powder compacts after the cold isostatic press (CIP) process; (**c**) transparent sintered bodies with ZrO_2 additives after hot isostatic press (HIP) treatment; (**d**) crushed sintered bodies without ZrO_2 additives after sintering.

The microstructures of the $(Tb_{0.6}Y_{0.4})_2O_3$ and Tb_2O_3 ceramics after HIP treatment were observed by TEM (transmission electron microscopy, ARM-200F, JOEL, Tokyo, Japan), and their images are shown in Figure 2a. It was confirmed that both ceramics were composed of grains of size of the order of several μm with different crystal orientations, and neither secondary phases nor grain boundary phases were observed. The lattice structure was observed to be a Bixbyite structure towards the grain boundary regions, and a clean grain boundary was confirmed. TEM-EDS (electron dispersive spectroscopy) analysis results of inner grain and grain boundary revealed that there was no segregation of ZrO_2, which was added as a sintering additive. Figure 2b shows the transmission polarized optical microscopic images of the $(Tb_{0.6}Y_{0.4})_2O_3$ and Tb_2O_3 ceramics. There was no birefringence, and they were optically homogeneous. No residual pores, the main factor in optical scattering, were detected inside the materials. XRD (X-ray diffraction, X'PERT PRO MPD, Malvern Panalytical, Almelo, The Netherlands) results revealed that the crystal system was only cubic phase and there were no other phases. When these ceramics were heat treated above 1400 °C, an orthorhombic ⇔ cubic phase transition occurred and the general optical quality was degraded. As seen in the above SEM and TEM images, the microstructures of the developed TYO ceramics were of a high-quality finished form.

It was anticipated that the added ZrO_2 would play an important role in inhibiting grain growth during the sintering of TYO ceramics; hence, the sintered bodies were composed of fine grains. Accordingly, (1) damage due to phase transition was effectively reduced by forming numerous grain boundaries, or (2) ZrO_2 itself possibly inhibited the phase transition. In the case of TYO ceramics without the addition of ZrO_2, the optical quality of the sample was very poor due to the significant grain growth during the sintering process or the sample having broken into pieces.

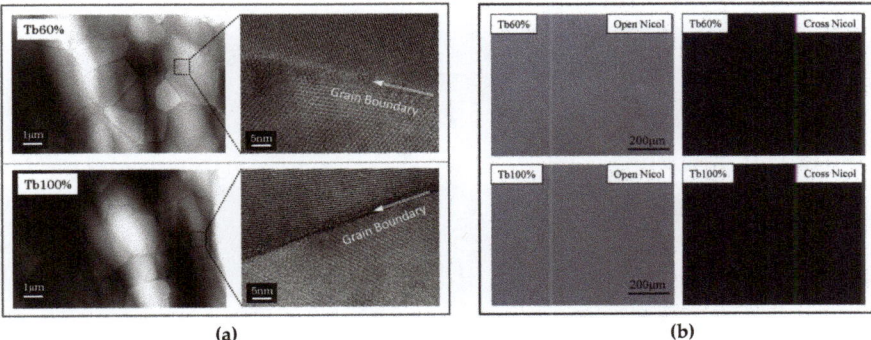

Figure 2. (a) The microstructure of $(Tb_{0.6}Y_{0.4})_2O_3$ and Tb_2O_3 ceramics after HIP treatment as observed by TEM. (b) Transmission polarized optical microscopic images of the produced $(Tb_{0.6}Y_{0.4})_2O_3$ and Tb_2O_3 ceramics.

3.2. Optical Properties of the Advanced Materials

The in-line transmittance curves of the $(Tb_{0.6}Y_{0.4})_2O_3$ (Tb 60%) and Tb_2O_3 (Tb 100%) ceramics from the visible to near-infrared wavelength regions are shown in Figure 3. The thickness of each sample was 5 mm, and the surfaces were optical polished but without AR (anti-reflective) coating. Only absorption due to Tb^{3+} ions can be confirmed around 480 nm, and the absorption of Tb 100% was stronger than that of Tb 60%. Wavelength dependency of the transmission lines was not detected for both ceramics, suggesting no Rayleigh scattering inside the materials. The refractive indices of Tb60% and Tb 100% materials at a 1μm wavelength are 1.920 and 1.940, respectively. The surface reflection loss (Fresnel loss) at a surface against air can be calculated by the following equation:

$$\beta(\lambda) = [(n(\lambda) - 1)^2]/[(n(\lambda) + 1)^2], \quad (1)$$

where $\beta(\lambda)$ is a reflection loss and n is a refractive index. Accounting for both sides of a sample, this Fresnel loss value is doubled to obtain theoretical transmittance. Their measured transmittance values, especially at a 1 μm wavelength, were very close to the theoretical values calculated by subtracting the Fresnel loss due to surface reflection. This result indicates that their optical losses are very low. An external view, polarized image, Schlieren image, and wavefront image by interferometry of the $(Tb_{0.6}Y_{0.4})_2O_3$ and Tb_2O_3 ceramics are shown in Figure 4a. The thickness of each sample was 11 mm. No optical inhomogeneity was observed in any inspection methods. The wavefront distortion was less than $\lambda/10$, suggesting that the ceramics are well suited for use as optical materials.

A laser with an output power of 30 mW (1064 nm wavelength, beam spot size: 2 mm) and with a TEM_{00} mode was used as a light source to evaluate the beam quality after passing through the sample. For comparison, a commercially available TGG single crystal was also measured as a reference. The original beam pattern and those after passing through the TGG single crystal and the produced $(Tb_{0.6}Y_{0.4})_2O_3$ and Tb_2O_3 ceramics are compared and summarized in Figure 4b. The original beam pattern was in the Gaussian mode, and the beam patterns that passed through the TGG single crystal, $(Tb_{0.6}Y_{0.4})_2O_3$, and Tb_2O_3 ceramics were almost unchanged. This result suggested that the variation of the refractive index inside the material is extremely small, which is in accordance with the measured results shown in Figure 4a.

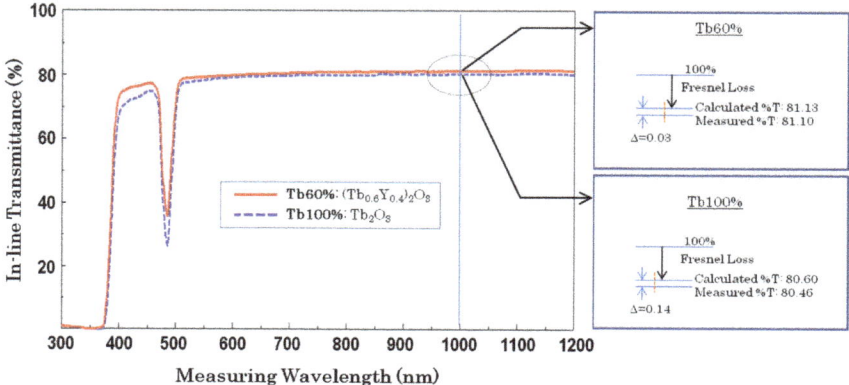

Figure 3. In-line transmittance curves of the $(Tb_{0.6}Y_{0.4})_2O_3$ and Tb_2O_3 ceramics (Sample thickness = 5 mm, optical-polished surfaces).

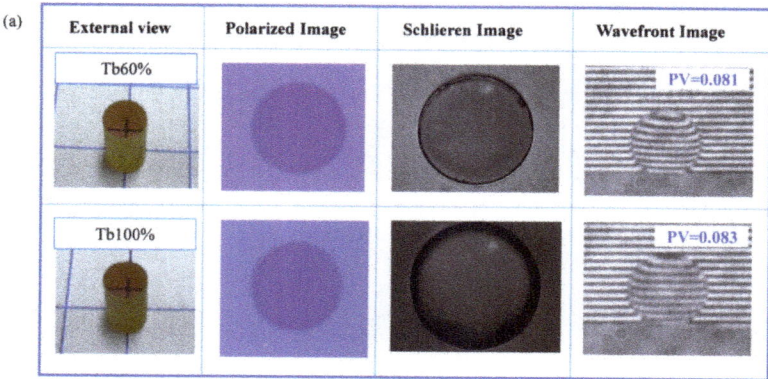

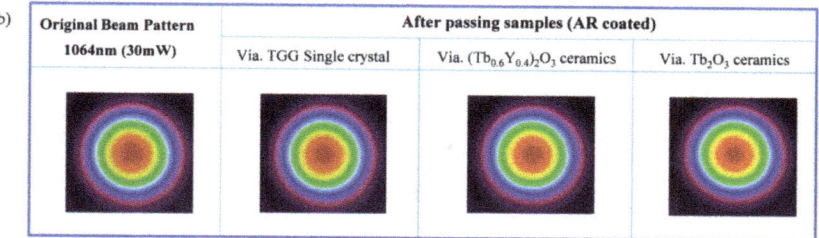

Figure 4. (**a**) External view, polarized image, Schlieren image, and wavefront image by interferometry of the $(Tb_{0.6}Y_{0.4})_2O_3$ and Tb_2O_3 ceramics. (The thickness of each sample was 11 mm.) (**b**) Original beam pattern and beam patterns after passing through the TGG single crystal and the produced $(Tb_{0.6}Y_{0.4})_2O_3$ and Tb_2O_3 ceramics.

When a laser with an output power of 50 W laser (1070 nm wavelength, CW (continuous wave) single-mode ytterbium fiber laser manufactured by IPG photonics corp., Burbach, Germany) was used as a light source, the beam shape after passing through the $(Tb_{0.6}Y_{0.4})_2O_3$ sample was slightly deformed due to the thermal lens effect ($1/f = 0.40$ m^{-1}: change in beam waist of passed laser beam). When the same measurement method was used, the value for TGG crystal was $1/f = 0.35$ m^{-1}, which is slightly better than that for the $(Tb_{0.6}Y_{0.4})_2O_3$ ceramics. However, for Y_2O_3 ceramics, which have similar optical loss, the value was $1/f = 0.34$ m^{-1}. As seen in Figure 1c, it is considered that a trace amount of Tb^{4+} ions remained in the $(Tb_{0.6}Y_{0.4})_2O_3$ ceramics, and optical absorption by these Tb^{4+} ions caused the thermal lens effect during laser irradiation. In addition, it was also confirmed that the material was not damaged by a power handling test which is generally used for commercial isolators developed for fiber lasers. In this test, a pulsed laser (pulse width 50 ps, peak power 0.3 MW, beam spot Φ 0.7 mm, power density 78 MW/cm^2) was irradiated at 2 MHz for 7000 h. In addition, when a laser damage test using an Nd:YAG laser with a wavelength of 1064 nm, a pulse width of 4 ms, and a laser focusing diameter of Φ 50 μm was performed, TYO $((Tb_{0.6}Y_{0.4})_2O_3)$ and TGG single crystal were damaged at an average power of 18 and 10 J/cm^2, respectively. From this result, it was also confirmed that the damage threshold of TYO ceramics is excellent, and the only technical issue is to reduce the thermal lens effect faintly generated during laser irradiation.

The wavelength dependencies of the Verdet constant for the TGG single crystal and the $(Tb_{0.6}Y_{0.4})_2O_3$ and Tb_2O_3 ceramics measured at wavelengths 633, 800, 980, 1030, and 1064 nm are shown in Figure 5a. At any wavelength, the Verdet constants of $(Tb_{0.6}Y_{0.4})_2O_3$ and Tb_2O_3 ceramics were about two times higher that of the TGG single crystal. In particular, in the case of the Tb_2O_3 ceramics, the Faraday rotation angle was found to be about 3.8 times (as a maximum) higher than that of the TGG single crystal. The relationship between the concentration of Tb ions (i.e., the occupancy of Tb^{3+} ions in the total number of cations) in TYO ceramics with a Bixbyite structure and the Verdet constant for 1 μm wavelength is shown in Figure 5b. The Verdet constants of Faraday rotator materials with other crystal structures are also plotted in the same figure as a reference. The Verdet constant simply increased with increasing Tb ion concentration, and 154 rad·T^{-1}·m^{-1} was achieved as the highest value. As the Verdet constant increases, the required length of the Faraday rotator can be reduced.

In the case of non-magnetic materials, the required length of the Faraday elements such as TGG single crystal and TYO ceramics to provide a 45° Faraday rotation angle depends on the strength of applied magnetic field. In the case of TGG single crystal, a length of 20 mm is required to obtain a 45° Faraday rotation angle when 1 T of magnetic field is applied. By using the ceramics with the highest Verdet constant, the length can be shortened to 5.1 mm to obtain a 45° Faraday rotation angle under 1 T of magnetic field, suggesting that the Faraday rotator device can be manufactured with a very compact size, whereas a commercial TGG crystal requires a length of about 20 mm. As seen in Figure 5b, TYO ceramics with a Bixbyite structure showed very high Verdet constants against Tb concentration (i.e., Tb ion number/total cation number) compared to other Faraday rotator materials such as TGG [13], TSAG [16], TAG [19], NTF ($Na_{0.37}Tb_{0.63}F_{2.26}$) [14,15], KTF ($KTb_3F_{10}$) [20], and TTO ($Tb_2Ti_2O_7$) [21].

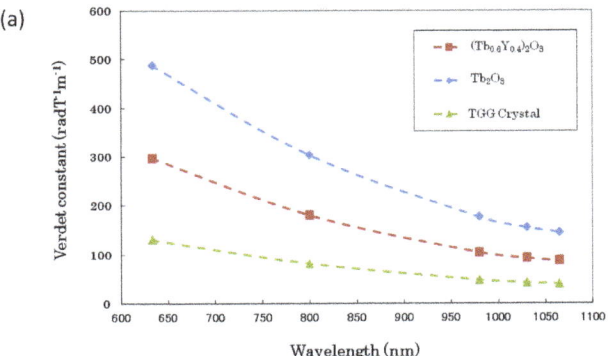

Figure 5. *Cont.*

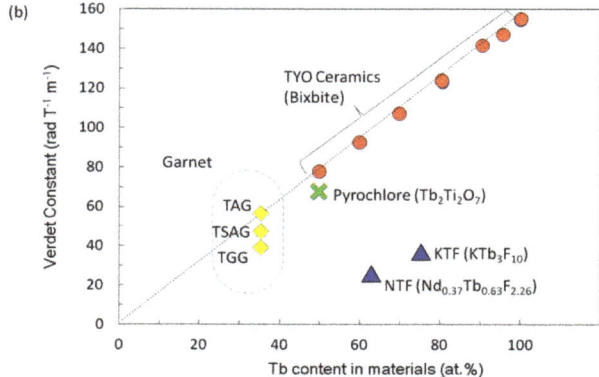

Figure 5. (**a**) Wavelength dependency of the Verdet constant for the TGG single crystal and the $(Tb_{0.6}Y_{0.4})_2O_3$ and Tb_2O_3 ceramics. (**b**) Relationship between the concentration of Tb ions in $(Tb_xY_{1-x})_2O_3$ ceramics and Verdet constant for 1 µm wavelength.

The measured results for transmitted laser power against the rotation angle of the polarizer under a 1 T magnetic field are shown in Figure 6. The Faraday rotation characteristics of the $(Tb_{0.6}Y_{0.4})_2O_3$ ceramics with very high Verdet constant were analogous to those of the commercial TGG single crystal, and their transmitted laser power values at a 45° rotation angle of the polarizer were comparable to each other (insertion loss: 0.04 dB). Compared with the maximum extinction ratio (E.R.) of the TGG single crystal (E.R.: 35 dB) where the rotation angle of polarizer was at −45°, the extinction ratio for the $(Tb_{0.6}Y_{0.4})_2O_3$ ceramics was as high as 42 dB. These materials with a very high Verdet constant can provide an equivalent Faraday rotation angle and a high extinction ratio even from a half-length (8.0 mm) of the conventional TGG single crystal (length: 20 mm). In other words, an advantage to replacing existing Faraday rotators with a high-Verdet-constant Faraday rotator is that if the length is kept the same size as the conventional single crystal, the required magnetic field can be reduced by more than half. In the case of Tb_2O_3 ceramics, principally it can be reduced to about one-fourth to obtain the same Faraday rotation performance as $(Tb_{0.6}Y_{0.4})_2O_3$ ceramics. The Tb_2O_3 ceramic (t = 5.1 mm), which had the highest Verdet constant, also showed similar behavior to TGG crystal, its insertion loss (I.L.) was 0.19 dB at a 45° rotation angle of the polarizer, and its extinction ratio was 47 dB at a −45° or 135° rotation angle of the polarizer.

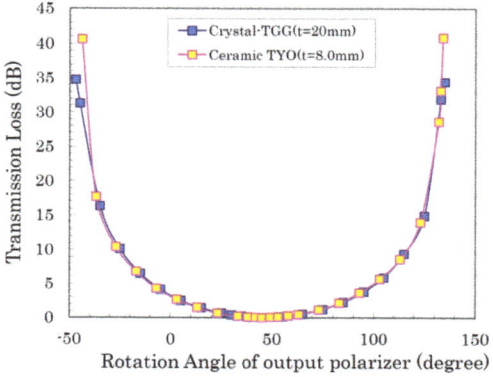

Figure 6. Faraday rotation characteristics of the TYO ceramics in comparison with those of the commercial TGG single crystal.

4. Discussion

The magneto-optical performance represented by the Verdet constant of Faraday rotator materials for the visible to 1.2 µm wavelength regions is not high enough since they do not include the magnetic element Fe in the composition of the materials. Although TAG [22–24] or TGG [13] single crystal are commonly used there, their Verdet constants are limited, and there is practically no other material which possesses a higher Verdet constant. It is very difficult to grow TAG single crystals by the CZ (Czochralski) method with acceptable aperture size because of their incongruent melting nature and unstable TAG phase in the Tb_2O_3-Al_2O_3 system [23]. The FZ (floating zone) method has often been used to grow TAG crystal, but its application is limited to research purposes only. Recently, it has been reported for TGG and TAG polycrystalline ceramics [25–29], but these do not exceed the conventional single-crystal Faraday rotator in terms of fundamental performance. The authors have demonstrated high-quality $(Tb_{1-x}Y_x)_3Al_5O_{12}$ (TAG) ceramics showing superior optical properties than the TAG single crystals prepared by the FZ method. It is the only example that has solved the problems relating to TAG single crystals and the performance of the conventional TGG single crystal.

In order to achieve a 45 degree Faraday rotation angle using a conventional single crystal under 1 T magnetic field, the rotator material needs to be as long as 20 mm; hence, a larger magnet is needed to supply a large enough magnetic field. Basically, the value of the Faraday rotation angle of the Faraday rotator is defined by the value of the Verdet constant, and the Verdet constant is determined dominantly by the occupancy of Tb ions which contributes to the magneto-optical effect, i.e., the Ga ions in TGG do not contribute to the Faraday rotation. TAG and TGG both have a garnet structure; hence, the occupancy of Tb^{3+} ions (Tb/(Tb+Al)) in the total number of cations is limited to 37.5%. Accordingly, garnet materials are not sufficient to achieve large Faraday rotation from the viewpoint of Tb^{3+} ion occupancy.

Based on the above viewpoint, Tb_2O_3 material may result in the highest occupancy of Tb^{3+} ions. However, since the melting point of Tb_2O_3 in its phase diagram is about 2300 °C and it has a phase transition from an orthorhombic ⇔ cubic crystal system near 1400 °C to an orthorhombic ⇔ hexagonal crystal system near 2100 °C [17,18], it is theoretically impossible to grow Tb_2O_3 single crystals by the conventional melt growth process. We confirmed that the polycrystalline TYO (including Tb_2O_3) ceramics have appropriate properties as an efficient optical isolator and also that the fabrication process is economically efficient. Hence, we filed a patent on the TYO ceramics in 2011 [30]. In 2015, light-yellow-colored Tb_2O_3 single crystals grown by using a $Li_6Tb(B_2O_3)_3$ system were reported with a low melting point flux around 1235–1160 °C, which is lower than the phase transition points [31]. However, the resulting crystal size was about 5 mm × 5 mm × 1 mm, which is too small for practical applications. Furthermore, optical quality was discussed there only by means of transmission curves, which are not enough to evaluate the opacity of thin film. The scattering loss of this material has been roughly estimated to be as large as around 10%/pass which requires a rotator length of 5.1 mm from the Verdet constant. In addition to this, it is not possible to control the crystal orientation, i.e., axis of easy magnetization, during crystal growth. Snetkov et al. synthesized Tb^{3+}:Y_2O_3 ceramics [32] but the transparency of their material is extremely poor and it cannot be used for optical applications. Although the basic characteristics of the TYO ceramics were reported in our previous paper [33], it is important to demonstrate the total performance parameters which are required to determine it to be a practical Faraday rotator material.

On the other hand, optical-grade YAG ($Y_3Al_5O_{12}$) and Sesquioxide ceramics denoted by Re_2O_3 (Re: lanthanide rare-earths) have been reported recently. Basically, a small amount of laser active elements is doped into the Re_2O_3 host materials (Re: Sc, Y, Lu) which do not have phase transition points up to 2000 °C. Then they are sintered at high temperature (over 1700 °C) to produce transparent ceramic materials. It has been confirmed that those materials are suited to laser gain media or scintillators, etc. [34,35]. It has been found that the Tb-containing materials with composition TYO can solve the technical difficulties of the conventional isolators. Here, it is necessary to increase the content of Tb ions in the solid solution to as high as possible. However, a technical issue there is that the fabrication temperature needs to be decreased with increasing Tb ion content in the solid solution. In particular, the ultimate material, Tb$_2$O$_3$, has very low phase transition temperature; hence, it is necessary to fabricate even Tb$_2$O$_3$ polycrystalline ceramics at temperatures lower than 1400 °C. In this report, we examined how we succeeded in densifying the materials to get high transparency, and finally examined the possibility of applications to optical isolators.

The question arises as to why single crystals have only been used as Faraday rotators so far. The reason for this is that optical scattering loss is small in single crystals and a uniform Faraday rotation angle can be achieved when a magnetic field is applied to a unidirectionally oriented crystal (<111> in general). It is well known that even a cubic crystal system has magneto-crystalline anisotropy. For example, in the case of TSAG (Terbium-Scandium-Aluminum Garnet) crystal, the Verdet constant of the <111> orientation, which has the smallest surface energy, is about 5–7% larger than that of the <110> or <100> orientation. [36] It is anticipated that the Verdet constants of other crystal orientations in a cubic system, which have higher surface energy, will be smaller than that of the <111> orientation. Consequently, it is likely that the Verdet constant of polycrystalline ceramics composed of numerous microcrystallites with randomized crystal orientations might be different from one grain to another, and the Faraday rotation behavior of polycrystalline ceramics might be different from that of single-crystal Faraday rotator materials. Here, the Faraday rotation angle is expressed as the equation below [37,38]:

$$\theta_F = VHL, \quad (2)$$

where θ_F is the Faraday rotation angle, V is the Verdet constant, H is the applied magnetic field, and L is the length of the Faraday rotator material.

In the case of ceramic materials, each grain has a randomized crystal orientation, and when a magnetic field is applied, the Faraday rotation angle will be slightly different in each grain of the ceramics; however, this difference did not cause any disadvantages upon utilization. When they were compared to the TGG single crystalline materials, it was confirmed that the TYO ceramics have equivalent values of the Faraday rotation angle and higher extinction ratios. In addition, this study has revealed the discovery of a novel Faraday rotator material which possesses a Verdet constant approximately 4 times higher than that of the commercial TGG single crystal.

5. Conclusions

Since the Faraday effect was discovered in 1845, a wide variety of Faraday rotator materials, such as glass or single crystal, have been developed and have progressed into the practical phase with developments in the field of telecommunication and machining. Only single-crystal materials have been put into practical use until now, but this does not necessarily mean that Faraday rotator materials are only limited to single crystals. As this work demonstrated, it is also possible to create new materials with excellent performance by other inorganic material processes such as ceramic fabrication processes. We successfully produced optical-grade TYO ceramics in this work, and their total performance as a Faraday rotator is summarized below.

(1) Optical-grade polycrystalline TYO ceramics with extremely low scattering were successfully produced for the first time.
(2) The Verdet constants of the TYO ceramics increased with increasing Tb concentration in the Bixbyite structure, and Tb$_2$O$_3$ showed the highest value: 3.8 times higher than that of the commercially available TGG single crystal.
(3) The Faraday rotation characteristics of the polycrystalline TYO ceramics were basically comparable to those of single-crystal isolator materials. In addition, one of the advantages was the possession of a large extinction ratio and a large Verdet constant, which can improve the performance of the isolator and downsize the device.

(4) The laser damage threshold of the TYO ceramics was as high as 18 J/cm^2 and they were resistant to pulsed laser damage (power density 78 MW/cm^2 and no damage during a 7000-hour durability test at 2 MHz).

(5) The thermal lens value, $1/f$ = 0.40 m^{-1}, of the TYO ceramics was slightly larger than that of TGG, probably due to a remaining trace amount of Tb^{4+} ions in the material. One of the remaining issues is to be able to use it for high-power and continuous-wave laser applications.

Supplementary Materials: The following are available online at http://www.mdpi.com/1996-1944/12/3/421/s1, Figure S1:(a) External view, and (b) polarized optical microscopic image of (Tb$_{0.5}$Y$_{0.5}$)$_2$O$_3$ single crystal; Figure S2: Relationships between the Tb ion concentration and the refractive index, and the thermal conductivity; Figure S3: (a) Prototype of optical isolator using TYO (Tb-60%) ceramic in comparison with commercial TGG optical isolator. (b) Schematic diagram of optical isolator. (c) Magnetic flux distribution inside the magnet house of optical isolator and the position of Faraday rotator sample influenced by the magnetic field. (d) Comparison of features of each Faraday rotator material; Figure S4: Appearance of large scaled TYO ceramic samples with various aperture sizes.

Author Contributions: Conceptualization, A.I., S.M.; methodology, A.I., Y.L.A., S.M.; validation, A.I., Y.L.A., S.M., A.Y.; formal analysis, A.I., Y.L.A., S.M., A.Y.; investigation, A.I., Y.L.A., S.M., A.Y.; resources, A.I., Y.L.A., S.M., A.Y.; writing—original draft preparation, A.I., Y.L.A.; writing—review and editing, A.I., Y.L.A., S.M., A.Y.; visualization, A.I., Y.L.A.; supervision, A.I., S.M.; project administration, A.I., S.M.

Funding: This research received no external funding.

Conflicts of Interest: The authors declare no conflict of interest.

References

1. Ikesue, A.; Kinoshita, T.; Kamata, K.; Yoshida, K. Fabrication and Optical Properties of High-Performance Polycrystalline Nd:YAG Ceramics for Solid-State Lasers. *J. Am. Ceram. Soc.* **1995**, *78*, 1033–1040. [CrossRef]
2. Sanghera, J.; Kim, W.; Villalobos, G.; Shaw, B.; Backer, C.; Frantz, J.; Sadowski, B.; Aggarwal, I. Ceramic Laser Materials. *Materials* **2012**, *5*, 258–277. [CrossRef] [PubMed]
3. Yamamoto, B.M.; Bhachu, B.S.; Cutter, K.P.; Fochs, S.N.; Letts, S.A.; Parks, C.W.; Rotter, M.D.; Soules, T.F. The Use of Large Transparent Ceramics in a High Powered, Diode Pumped Solid State Laser. In Proceedings of the Advanced Solid-State Photonics, Nara, Japan, 27–30 January 2008; p. WC5.
4. Tokurakawa, M.; Takaichi, K.; Shirakawa, A.; Ueda, K.; Yagi, H.; Hosokawa, S.; Yanagitani, T.; Kaminskii, A.A. Diode-pumped mode-locked Yb^{3+}:Lu$_2$O$_3$ Ceramic Laser. *Opt. Express* **2006**, *14*, 12832–12838. [CrossRef] [PubMed]
5. Young, A.T. Rayleigh scattering. *Appl. Opt.* **1981**, *20*, 522–535. [CrossRef]
6. Strutt, J. On the scattering of light by small particles. *Philos. Mag.* **1871**, *41*, 447–454. [CrossRef]
7. Ikesue, A.; Aung, Y.; Taira, T.; Kamimura, T.; Yoshida, K.; Messing, G. Progress in Ceramics Lasers. *Annu. Rev. Mater. Res.* **2006**, *36*, 397–429. [CrossRef]
8. Kong, L.B.; Huang, Y.; Que, W.; Zhang, T.; Li, S.; Zhang, J.; Dong, Z.; Tang, D. Transparent Ceramics. In *Mining, Metallurgy and Material Engineering*; Bergmann, C.P., Ed.; Springer: Berlin/Heidelberg, Germany, 2015.
9. Ikesue, A.; Aung, Y.L. Ceramic Laser Materials. *Nat. Photonics* **2008**, *21*, 721–726. [CrossRef]
10. Tamaki, T.; Kaneda, H.; Kawamura, N. Magnet-optical properties of terbium bismuth iron oxide ((TbBi)$_3$Fe$_5$O$_{12}$) and its application to a 1.5 μm wideband optical isolator. *J. Appl. Phys.* **1991**, *70*, 4581–4583. [CrossRef]
11. Zhang, G.Y.; Xu, X.W.; Chong, T.C. Faraday rotation spectra of bismuth-substituted rare-earth iron garnet crystals in optical communication band. *J. Appl. Phys.* **2004**, *95*, 5267–5270. [CrossRef]
12. Khazanov, E.; Andreev, N.; Palashov, O.; Poteomkin, A.; Sergeev, A.; Mehl, O.; Reitze, D.H. Effect of terbium gallium garnet crystal orientation on the isolation ratio of a Faraday isolator at High Average Power. *Appl. Opt.* **2002**, *41*, 483–492. [CrossRef]
13. Barnes, N.P.; Petway, L.B. Variation of the Verdet constant with temperature of terbium gallium garnet. *J. Opt. Soc. Am. B* **1992**, *9*, 1912–1915. [CrossRef]
14. Mironov, E.A.; Palashov, O.V.; Voitovich, A.V.; Karimov, D.N.; Ivanov, I.A. Investigation of Thermo-Optical Characteristics of Magneto-Active Crystal Na$_{0.37}$Tb$_{0.63}$F$_{2.26}$. *Opt. Lett.* **2015**, *40*, 4919–4922. [CrossRef] [PubMed]
15. Karimov, D.N.; Sobolev, B.P.; Ivanov, I.A.; Kanorsky, S.I.; Masalov, A.V. Growth and Magneto-Optical Properties of Na$_{0.37}$Tb$_{0.63}$F$_{2.26}$ Cubic Single Crystal. *Crystallogr. Rep.* **2014**, *59*, 718–723. [CrossRef]
16. Yasuhara, R.; Snetkov, I.; Starobor, A.; Mironov, E.; Palashov, O. Faraday rotator based on TSAG crystal with <001> orientation. *Opt. Express* **2016**, *24*, 15486–15493. [CrossRef]

17. Adachi, G. Physics and Chemistry of Yttrium Compounds. *Bull. Ceram. Soc. Jpn.* **1988**, *23*, 430–437.
18. Coutures, J.P.; Vegers, R.; Foex, M. Comparison of solidification temperatures of different rare earth sesquioxides; effect of atmosphere. *Rev. Int. Hautes Temp. Refract.* **1975**, *12*, 181–185.
19. Aung, Y.L.; Ikesue, A. Development of optical grade $(Tb_xY_{1-x})_3Al_5O_{12}$ ceramics as Faraday rotator material. *J. Am. Ceram. Soc.* **2017**, *100*, 4081–4087. [CrossRef]
20. Stevens, K.T.; Schlichting, W.; Foundos, G.; Payne, A.; Rogers, E. Promising materials for high power laser isolators. *Laser Tech. J.* **2016**, *3*, 18–21. [CrossRef]
21. Guo, F.; Sun, Y.; Yang, X.; Chen, X.; Zhao, B.; Zhuang, N.; Chen, J. Growth, Faraday and inverse Faraday characteristics of $Tb_2Ti_2O_7$ crystal. *Opt. Express* **2016**, *24*, 5734–5743. [CrossRef]
22. Yoshikawa, A.; Kagamitani, Y.; Pawlak, D.A.; Sto, H.; Machida, H.; Fukuda, T. Czochralski Growth of $Tb_3Sc_2Al_3O_{12}$ Single Crystal for Faraday Rotator. *Mater. Res. Bull.* **2002**, *37*, 1–10. [CrossRef]
23. Geho, M.; Takagi, T.; Chiku, S.; Fujii, T. Development of Optical Isolators for visible light using Terbium Aluminum Garnet ($Tb_3Al_5O_{12}$) Single Crystals. *Jpn. J. Appl. Phys.* **2005**, *44*, 4967–4970. [CrossRef]
24. Ganschow, S.; Klimm, D.; Reiche, P.; Uecker, R. On the Crystallization of Terbium Aluminum Garnet. *Cryst. Technol.* **1999**, *34*, 615–619. [CrossRef]
25. Yoshida, H.; Tsubakimoto, K.; Fujimoto, Y.; Mikami, K.; Fujita, H.; Miyanaga, N.; Nozawa, H.; Yagi, H.; Yanaggitani, T.; Nagaya, Y.; et al. Optical Properties and Faraday Effect of Ceramic Terbium Gallium Garnet for a Room Temperature Faraday Rotator. *Opt. Express* **2011**, *19*, 15181–15197. [CrossRef] [PubMed]
26. Yasuhara, R.; Tokita, S.; Kawanaka, J.; Kawashima, T.; Kan, H.; Yagi, H.; Nozawa, H.; Yanagitani, T.; Fujimoto, Y.; Yoshida, H.; et al. Cryogenic temperature characteristics of Verdet constant on terbium gallium garnet ceramics. *Opt. Express* **2007**, *15*, 11255–11261. [CrossRef] [PubMed]
27. Yasuhara, R.; Snetkov, I.; Starobor, A.; Zheleznov, D.; Palashoz, O.; Khazanov, E.; Yanagitani, T. TGG Ceramics Faraday Rotator for High Power Laser Application. *Opt. Lett.* **2014**, *39*, 1145. [CrossRef] [PubMed]
28. Zheleznov, D.; Atarobor, A.; Palashov, O.; Chen, C.; Zhou, S. High Power Faraday Isolators based on TAG Ceramics. *Opt. Express* **2014**, *22*, 2578–2583. [CrossRef] [PubMed]
29. Zheleznov, D.; Atarobor, A.; Palashov, O.; Lin, H.; Zhou, S. Improving Characteristics of Faraday Isolator based on TAG Ceramics by Cerium Doping. *Opt. Lett.* **2014**, *39*, 2183–2186. [CrossRef]
30. Makikawa, S.; Yahagi, A.; Ikesue, A. Transparent Ceramic, Method for Manufacturing Same, and Magneto-Optical Device. U.S. Patent 9,470,915, 10 October 2016.
31. Veber, P.; Velazquez, M.; Gardet, G.; Rytz, D.; Peltz, M.; Decourt, R. Fluxgrowth at 1230 °C of Cubic Tb_2O_3 Single Crystals and Characterization of their Optical and Magnetic Properties. *Cryst. Eng. Commun.* **2015**, *17*, 492–497. [CrossRef]
32. Snetkov, I.L.; Permin, D.A.; Balabanov, S.S.; Palashov, O.V. Wavelength Dependence of Verdet constant of $Tb^{3+}:Y_2O_3$ Ceramics. *J. Appl. Phys. Lett.* **2016**, *108*, 161905. [CrossRef]
33. Ikesue, A.; Aung, Y.L.; Makikawa, S.; Yahagi, A. Polycrystalline $(Tb_xY_{1-x})_2O_3$ Faraday Rotator. *Opt. Lett.* **2017**, *42*, 4399–4401. [CrossRef]
34. Wang, L.; Huang, H.; Shen, D.; Zhang, J.; Chen, H.; Wang, Y.; Liu, X.; Tang, D. Room Temperature continuous-wave Laser Performance of LD pumped $Er:Lu_2O_3$ and $Er:Y_2O_3$ Ceramics at 2.7 μm. *Opt. Exp.* **2014**, *22*, 19495–19503. [CrossRef] [PubMed]
35. Newburgh, G.A.; Word-Daniels, A.; Michael, A.; Merkle, L.D.; Dubinskii, A.I.M. Resonantly Diode-Pumped $Ho^{3+}:Y_2O_3$ Ceramic 2.1 μm Laser. *Opt. Express* **2011**, *19*, 3604. [CrossRef] [PubMed]
36. Kagamitani, Y.; Pawlak, D.A.; Sato, H.; Yoshikawa, A.; Martinek, J.; Machhida, H.; Fukuda, T. Dependence of Faraday Effect on the Orientation of Terbium- Scandium- Aluminum Garnet Single Crystal. *J. Mater. Res.* **2004**, *19*, 579–583. [CrossRef]
37. Kohli, J.T. Volume 67 Ceramic Transaction. In *Faraday Effect in Lanthanide-Doped Oxide Glasses*; American Ceramic Society: Westerville, OH, USA, 1995; pp. 125–136, ISBN 1-57498-012-2.
38. Stadler, B.J.H.; Vaccaro, K.; Davis, A.; Martin, E.A.; Lorenzo, J.P. Volume 60 Ceramic Transaction. In *Characterization of Magneto-Optical Mn-Doped InGaAsP Thin Films on InP*; American Ceramic Society: Westerville, OH, USA, 1995; pp. 195–204, ISBN 1-57498-003-3.

© 2019 by the authors. Licensee MDPI, Basel, Switzerland. This article is an open access article distributed under the terms and conditions of the Creative Commons Attribution (CC BY) license (http://creativecommons.org/licenses/by/4.0/).

Review

Holography with Photochromic Diarylethenes

Luca Oggioni [1,2], Giorgio Pariani [1], Frédéric Zamkotsian [3], Chiara Bertarelli [2] and Andrea Bianco [1,*]

1. INAF-Osservatorio Astronomico di Brera, via E. Bianchi 46, 23807 Merate (LC), Italy
2. Politecnico di Milano, Dipartimento di Chimica, Materiali e Ingegneria Chimica 'Giulio Natta', P.zza L. da Vinci 32, 20133 Milano (MI), Italy
3. Aix Marseille Universite, CNRS, CNES, LAM, Laboratoire d'Astrophysique de Marseille, 38 Rue Frédéric Joliot Curie, 13388 Marseille CEDEX 13, France
* Correspondence: andrea.bianco@inaf.it; Tel.: +39-0272320460

Received: 22 July 2019; Accepted: 26 August 2019; Published: 1 September 2019

Abstract: Photochromic materials are attractive for the development of holograms for different reasons: they show a modulation of the complex refractive index, meaning they are suitable for both amplitude and phase holograms; they are self-developing materials, which do not require any chemical process after the light exposure to obtain the final hologram; the holograms are rewritable, making the system a convenient reconfigurable platform for these types of diffractive elements. In this paper, we will show the features of photochromic materials, in particular diarylethenes in terms of the modulation of a transparency and refractive index, which are mandatory for their use in holography. Moreover, we report on the strategies used to write binary and grayscale holograms and their achieved results. The outcomes are general, and they can be further applied to other classes of photochromic materials in order to optimize the system for achieving high efficiency and high fidelity holograms.

Keywords: holography; photochromism; diarylethenes; refractive index; CGH

1. Introduction

The possibility of storing a 3D scene in a substrate has been a dream for a long time. Thanks to Gabor and his invention of holography in 1948 [1] and laser development in the following decades [2], such a dream has come true. Since then, holography has found many potential technological uses, while important developments for both theory and application have been achieved [3].

When considering the hologram manufacturing, issues related to photosensitive material are crucial. Indeed, an ideal material for hologram manufacturing should show [4]: a high spatial resolution, a large dynamic range, a good signal to noise ratio, high optical quality, and large sensitivity in a wide spectral range. Another attractive property that holographic materials may show is the ability to self-develop, namely, no chemical process is required after the pattern transfer to obtain the final usable hologram. Clearly, the choice of the photosensitive material depends on different factors, in particular, if the hologram is a phase or amplitude, and the technique used to transfer the pattern. In addition, strategies to obtain holograms that are reconfigurable and switchable are highly desired.

There are different approaches to achieve these kinds of diffractive devices and photochromic dyes surely are an interesting option. Nice features of such materials include their rewritability, which is intrinsic in the reversible transformation. Moreover, they can be used for making both amplitude and phase holograms [5]. Among the different classes of photochromic materials, T-type materials are interesting in the case of real-time holography because of their efficient thermal decoloration process [6]; whereas the P-type (thermally stable) holograms are much more interesting where re-addressable holograms are required. Diarylethenes are surely the most studied holograms for holographic optical memories, 3D displays, and holographic gratings [7–12] belonging to the P-type class, thanks to their well-known good overall photochromic properties [13] and the possibility of obtaining highly responsive films. In addition, the use of diarylethenes in combination with nanoparticles (in particular gold ones) could be of great interest in this field, since the optical properties and their switching can be tuned by acting both on the nanoparticles side (mainly size and dispersion, which affect the plasma frequency) and the photochromic unit side [14–19]. The performances of diarylethene based holograms are strongly related to the optimization of the photochromic substrate and to the writing procedure.

Other strategies and materials are possible to use to obtain reconfigurable holograms: photorefractive materials and photosensitive liquid crystals are two interesting families. As for photorefractive materials, they show a refractive index modulation as the result of the photoconductive and the Pockels effects [20], which makes them suitable for phase holographic elements [21]. Fast reconfigurable holograms [22,23], 3D holographic displays [24,25], and holographic memories [26] can be obtained thanks to the rapid growth of refractive index modulation, a very peculiar characteristic of such systems. On the other hand, the hologram is not usually persistent, so this approach is not suitable for long lasting devices. Regarding holographic liquid crystals (LCs), there are different possibilities since the LCs can change their properties (orientation, phase separation, and refractive index modulation) through both optical stimuli and electric stimuli. Consequently, rewritable systems, ON-OFF grating, and polarized sensitive gratings are possible. In the case of light sensitive LCs, azobenzene photochromic moieties are often used [27,28] and rewritable holograms can be obtained by achieving major modulation of the refractive index. Similar systems were also considered for making holographic memories [29]. By using polymer-dispersed liquid crystal combined with holography (H-PDLC), it is possible to obtain switchable phase gratings and other optical elements thanks to the phase separation that induces the refractive index modulation and the application of the electric field [30,31].

In this paper, we report on the main features of photochromic diarylethenes in terms of relevant properties for phase and amplitude holograms. A hybrid computation tool is shown to help the optimization of the films, mainly focusing on the chemical structure of the diaryelethene; the different strategies for writing photochromic holograms are also discussed. Examples are reported in order to support the discussion. The results here reported can be easily generalized to other classes of photochromic materials and could inspire the development of new/optimized photochromic systems for high efficiency and high fidelity holographic optical elements.

2. Computer-Generated Holograms

Holograms digitally calculated are called Computer-Generated Holograms (CGHs). The ideal wavefront to be reconstructed is computed on the basis of the diffraction theory, starting from the wave field distribution of the object beam [32,33]. Such an approach is of great interest since it allows for recording holograms of virtually any object or scene without the existence of the physical object. They can also use optical elements and filters to manipulate light phase and intensity. In the same manner as traditional holograms, CGHs are classified as either phase or amplitude.

In the scalar diffraction approach, if we neglect the reconstruction noise, depending on the hologram type and discretization levels, both types of hologram are able to reconstruct the desired object with the main difference being in the diffraction efficiency [32], as will be discussed later on (see Section 2.2). When diffraction efficiency is not an issue, amplitude holograms may be preferred for their easier manufacture. CGHs, thanks to their ability to generate custom wavefronts, are finding applications in beam shaping, particles manipulation, interferometric optical testing, and anti-counterfeiting [34–36].

Concerning interferometry, great efforts have been done in the recent years to improve CGH capabilities beyond their first development by Wyant almost 50 years ago [37]. Nowadays, they are used as optical surface references, to cope with the production of complex and non-standard optical surfaces (aspherical and free form), which are made possible by new optical fabrication technologies [38]. They are applied to test different optics, even large aspherical mirrors for the new generation telescopes [39,40].

Two main type of holograms can be recorded in the holographic material [41]: the Fourier hologram, which exploits the inverse Fourier transformation of the image, and the Fresnel hologram, which encodes the interference pattern of the wave propagated to the object.

2.1. Fourier and Fresnel CGHs

A collimated beam passing through a lens undergoes a Fourier transformation. Thus, starting from the image to be reconstructed, it is possible to calculate the complex wavefront to be encoded in the CGH using Equation (1). Moreover, the inverse transformation, can be used to reconstruct the image from the CGH pattern [42]. Here, we report the mathematical operator of the direct $\mathcal{F}$ and inverse $\mathcal{F}^{-1}$ Fourier Transformations:

$$g(\mu,v) = \mathcal{F}[f](\mu,v) = \iint_{R^2} f(x,y)\, e^{-2\pi i (x\mu + yv)}\, dx\, dy \tag{1}$$

$$f(x,y) = \mathcal{F}^{-1}[g](x,y) = \iint_{R^2} g(\mu,v)\, e^{2\pi i (x\mu + yv)}\, d\mu\, dv \tag{2}$$

where x, y and μ, v are the coordinates of the image plane and Fourier space, respectively.

Fresnel holograms are directly calculated by propagating the wavefront to be reconstructed and exploiting the light propagation equations that are modeled by the Rayleigh-Sommerfeld diffraction theory [43]. If we consider to have the hologram plane in $z = 0$ and the object plane at z, we can write:

$$E_z(x,y) = \iint E_0(u,v)\, \frac{e^{ikr}}{r}\, du\, dv \tag{3}$$

The resulting complex wave E_z is estimated by calculating the sum of the contributions of each pixel of the hologram anywhere on the screen located at a distance z. Each pixel is considered as a secondary spherical wave source weighted by the function $E_0(u,v)$. These secondary waves are generated when the incident wave, characterized by its complex amplitude E_0 and wavelength λ, reaches the hologram. The same strategy can be applied to calculate E_0 by inverting Equation (3) and positioning the object plane at $-z$, in order to keep the same direction of propagation.

$$E_0(u,v) = \iint E_{-z}(x,y)\, \frac{e^{ikr}}{r}\, dx\, dy \tag{4}$$

The function $E_0(u,v)$ is the complex amplitude of the hologram, which must be approximated before the encoding.

2.2. Diffraction Efficiency

Once the complex electric field function at the hologram plane has been calculated, the next step is the encoding into the CGH. However, to transfer all the complex information, a material able to modulate both amplitude and phase of a wavefront is needed. Despite some successful attempts, multi-step processes and complex procedures are ultimately required for this approach [44]. The traditional approach is to code the complex wavefront in the form of a phase only or amplitude only map. The main difference between the two coding strategies is the hologram diffraction efficiency.

In order to give an estimation of the hologram efficiency, we make use of a model reported by Brown in 1969 [32], which considers monodimensional gratings with a periodical structure that is either binary or grayscaled. The wavelength of the incident light is assumed to be much smaller than the grating period, so the scalar diffraction approximation can be applied. Figure 1 reports the theoretical efficiency of gray-scaled (a) and binary (b) amplitude holograms, and gray-scaled (c), binary (d) and blazed (e) phase holograms. The efficiency is related to the amplitude of the modulation A, which varies between 0 and 1 in the case of amplitude holograms and between 0 and 2π in the case of phase holograms.

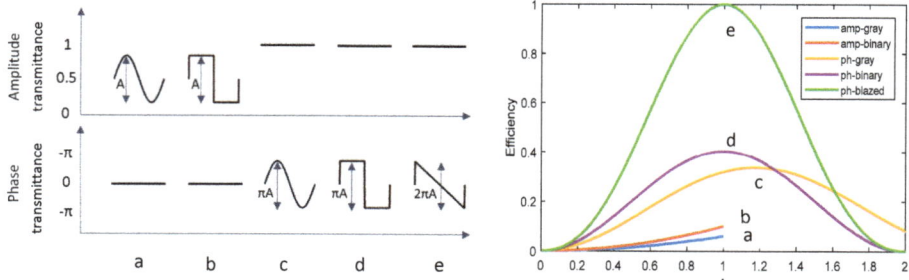

Figure 1. Modulation profiles for grayscaled (**a**), binary (**b**) amplitude holograms, grayscaled (**c**), binary (**d**) and blazed (**e**) phase holograms. Dependence of the first order diffraction efficiency on the modulation parameter A (right).

The efficiency of (a) and (b) reaches, in the best conditions, 6.2% and 10.1%, respectively, while for phase hologram (c) and (d) the efficiency is 34% and 41% respectively. In the case of the blazed hologram, it can even reach the 100% efficiency.

In the case of photochromic materials, both amplitude and phase holograms are possible, according to the working region of the hologram. For amplitude holograms, the parameter A is directly linked to the contrast between transparent and opaque regions, i.e., to the transmission of the photochromic film in the transparent and colored forms. The contrast, which is a wavelength dependent quantity, in a region where only one of the two forms is fully transparent, is given by the dye concentration C, the molar absorbance ε of the colored form, and the thickness of the film d as follows:

$$Contrast = \frac{T_{transparent}}{T_{coloured}} = \frac{1}{10^{-Abs}} = 10^{\varepsilon Cd} \tag{5}$$

Figure 2 reports the first order diffraction efficiency of a binary hologram as function of the film contrast [45].

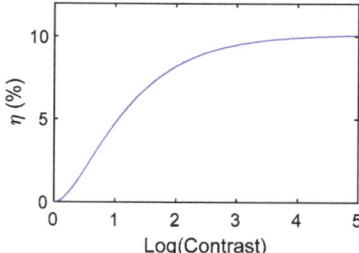

Figure 2. First order diffraction efficiency of a binary amplitude grating as function of the contrast value [45].

The contrast is asymptotic to the maximum efficiency of 10.1% for values larger than 5000, but values around 8% are good enough for many applications and in particular for interferometric purposes. These efficiencies are reached when the contrast is larger than 100, value that is obtained for optical density of the film in the colored form larger than two (considering again an absorbance zero for the uncolored form).

Concerning phase holograms, the key parameter to deal with is the product between the refractive index modulation and the film thickness $d \cdot \Delta n$. For example, if we take the efficiency of volume phase gratings working in the Bragg regime, we can write the first order diffraction efficiency at the Bragg angle (α_B) as [46]:

$$\eta = \frac{1}{2}\sin^2\left(\frac{\pi \Delta n d}{2\lambda \cos\alpha_b}\right) + \frac{1}{2}\sin^2\left(\frac{\pi \Delta n d}{2\lambda \cos\alpha_b}\cos(2\alpha_B)\right) \tag{6}$$

where λ is the wavelength of the incident light. We studied the efficiency dependence by Δn, considering $\lambda = 650$ nm, 750 nm, 850 nm and $\alpha_B = 19°$. In Figure 3, we report the results for a Δn in the range 0–0.08.

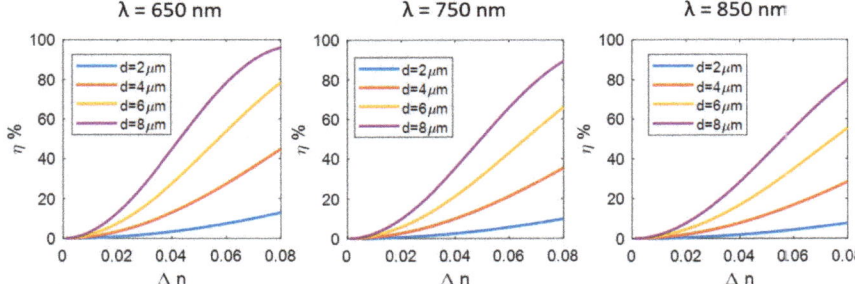

Figure 3. Theoretical efficiency of phase gratings as function of the refractive index modulation for films with a thickness of 2, 4, 6 and 8 µm. The results are shown for three different wavelength 650, 750 and 850 nm; the grating line density is 1000, 870, 770 lines/mm, respectively.

Considering that diarylethene based photochromic films can reach Δn of 1–4%, we are able to write phase binary gratings with good efficiency, depending on the film thickness. However, the maximum useful thickness of the photochromic materials is limited by the UV penetration (more details are provided later on), which determine the degree of conversion through the film thickness. Therefore, we can conclude that there is a sort of upper limit in the $d \cdot \Delta n$ value for the photochromic films.

3. Diarylethenes: Properties Modulation

Diarylethenes show a light-induced transformation between two forms a and b as reported in Figure 4 in the case of the perfluorocycplopentene derivatives. The a form, called open form, is usually uncolored since the π-conjugation is interrupted between the two side parts of the molecule. Upon illumination with UV light, a 4n + 2 electrocyclization occurs, and the b form, called close form, is obtained. This state is characterized by a π-conjugation extended along the whole molecular backbone, with a consequent coloration of the materials.

Figure 4. Photoreaction in 1,2-diarylethenes considered in this work. (**a**) open form (uncolored); (**b**) close form (colored). The detailed structures are reported in Figure 5.

Actually, hundreds of diarylethenes have been synthesized so far, and comprehensive reviews report on the main characteristics and possible applications of this important family of photochromic compounds [47,48]. In this review, we limit the discussion to a series of diarylethenes, and we discuss the modulation of absorption properties in the UV-Vis as a function of the chemical structure specifically to later highlight the conversion in the film state, which is of fundamental relevancy to reaching an adequate contrast in amplitude holograms. Moreover, for possible application as phase holograms, features for maximizing the refractive index modulation are reported.

3.1. UV-vis Absorption

In a liquid solution, the photochromic process approximately occurs uniformly in the whole volume, and the conversion at the photosteady state depends on the absorption coefficients (ε_A, ε_B) of the two isomeric forms at the irradiation wavelength and on the quantum yield of forward and backward reactions (ϕ_{AB}, ϕ_{BA}). All these quantities depend on the molecular building blocks, both those ones involved in the 4n + 2 electrocyclization (i.e., the photoactive part of the molecule) and the lateral substituents. Many diarylethene derivatives have been synthesized so far, and the effect on the specific chemical structure on the absorption properties for a selection of compounds (see Figure 5) is highlighted in Table 1.

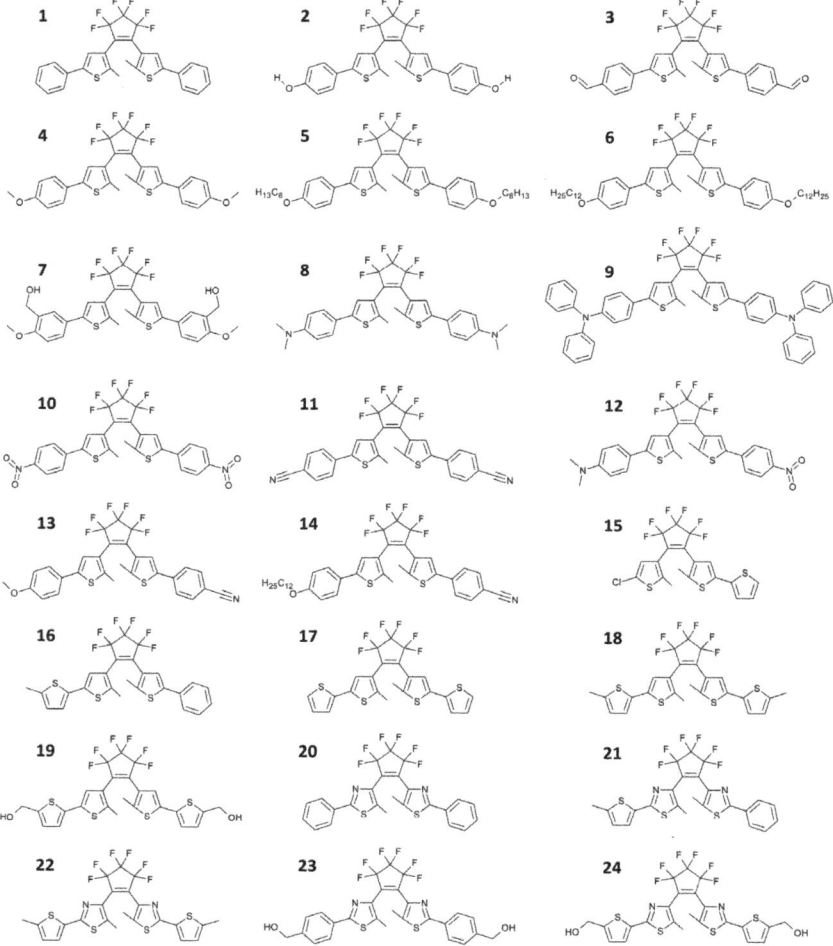

Figure 5. Series of 1,2-diarylethenes in their open state (uncolored). They differ for the aromatic ring in the switching structure (thienyl from **1** to **19** or thiazolyl from **20** to **24**) and for the lateral groups. Electroactive substituents can be also present to give push-push structures (compounds from **4** to **9**), pull-pull structure (compounds **3**, **10** and **11**) or push-pull structures (compounds from **12** to **14**) in their closed (colored) state.

Table 1. Absorption maxima (λ) and molar extinction coefficient (ε) of diaryletenes shown in Figure 5 in their two isomeric forms (uncolored and colored). Absorption spectra measured in hexane solution (* in EtOH).

Compound	Uncolored Form		Colored Form	
	λ_{UV} (nm)	ε_{UV} (M^{-1} cm^{-1})	λ_{VIS} (nm)	ε_{VIS} (M^{-1} cm^{-1})
1	278	34,660	575	14,990
2 *	288	38,720	584	16,400
3	328	40,450	608	14,450
4	293	43,610	582	20,050
5	297	41,660	583	21,110
6	294	36,580	582	16,940
7	296	35,980	592	18,010
8	320	49,960	607	24,300
9	353	51,040	613	28,100
10	315	27,670	598	9790
11	315	33,060	591	11,425
12	330	44,230	642	25,860
13	305	37,970	597	18,070
14	305	35,390	602	16,620
15	251	21,410	546	10,910
16	282	30,780	587	15,420
17	310	28,190	592	16,750
18	318	32,480	598	21,490
19 *	320	32,420	608	19,090
20	301	30,640	519	10,810
21	316	23,700	534	11,800
22	329	37,310	549	19,400
23 *	304	35,050	530	12,690
24 *	330	31,520	553	15,820

Despite the fact that the values reported in Table 1 seem to be highly scattered, correlations between the different parameters can be found for the different groups of diaryethenes herein synthesized and analyzed. The comparison between the molar extinction coefficients of the uncolored and the colored forms of any diarylethene (ε_{UV}, ε_{VIS}, respectively) shows that the maximum absorbance of the visible band of the colored isomer is roughly half of the absorbance in the UV (Figure 6a). Moreover, diarylethenes with lateral substituents characterized by the presence of a phenyl group, either alone or linked with a withdrawing functional group, have a lower intensity of the visible band (green series). Indeed, all of these molecules have a less-conjugated structure in their colored forms. Conversely, molecules belonging to the blue series have a more intense visible absorption, which can arise from all the possible different chemical structures allowing for an extended π-conjugation in the closed form, e.g., the use of thiophene-thiophene as lateral substituent (compounds from **17** to **19**) and the push-pull substituents (compounds from **12** to **14**). In addition, the triphenylamine as substituent is known to give an effective π-conjugation (herein compound **9**).

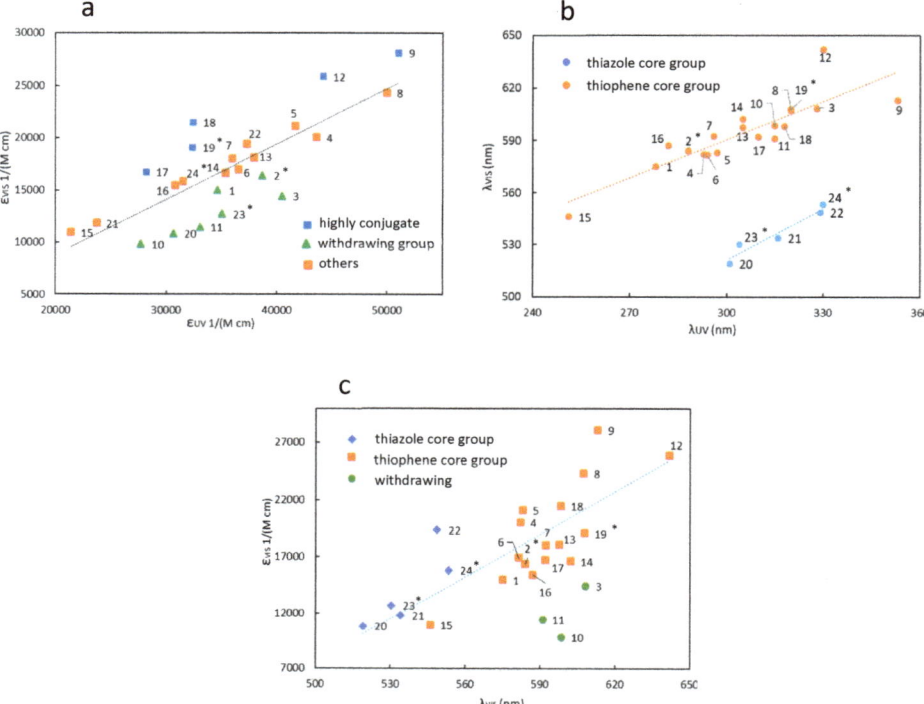

Figure 6. (a) Plot of the molar extinction at the maximum absorbance in the visible of the colored isomer ε_{vis} and in the UV of the transparent isomer ε_{UV}, for the diarylethenes of 5; (b) Plot of the wavelength of the visible peak for the colored isomer and for the UV peak of the transparent isomer; (c) Plot of the position and the molar extinction coefficient of the band in the visible for the colored isomers. Molecules were characterized in hexane except for * which were dissolved in ethanol.

The analogous analysis, but considering the absorption maxima instead of the absorption coefficient, leads to the general conclusion that a redshift of the visible band of the colored form corresponds to a redshift of the UV band of the uncolored form (Figure 6b). The two series of data in the figure correspond to the dithienylethenes (orange data) and the dithiazolylethenes (blue data). For both series, the wavelength gap between the absorption maxima of the colored and uncolored forms is approximately the same inside the members of the same series. In particular, the gap is about 220 nm for the thiazole based series and 290 nm for the thienyl based one. Actually, the presence of electroactive substituents can modify this wavelength gap, i.e., push-pull substituted dithienylethenes (compounds from **12** to **14**) are characterized by a larger λ_{VIS}-λ_{UV}, with the largest difference value for the compound **12**, having both very strong donor and acceptor groups.

Finally, the relationship between ε_{vis} and λ_{VIS} is highlighted (Figure 6c), since the behavior of the photochromic molecules in the visible (i.e., the contrast) is relevant for the development of amplitude holograms. The overall evidence is that the longer the wavelength of the peak, the higher its absorption intensity [49], which is a common trend in conjugated molecules [50]. However, the presence of withdrawing groups (e.g., compounds **9** and **10**) decreases the absorption coefficient, whereas donor groups lead to higher absorbance (compounds **4–9**).

At the solid state, including dyes in the crystalline or amorphous state and polymer dispersed dyes, the situation is more complex since the light-induced process proceeds from the outer layer to the inner layer. Actually, the full transformation from the colored to the uncolored forms is always possible, since only the colored isomer absorbs in the visible. Instead, the coloration process is not straightforward as both the isomeric states of diarylethenes absorb UV light. Therefore, the radiation is attenuated through the volume while the coloration proceeds and a limit depth of UV penetration exists, beyond which the photochromic reaction cannot further occur. In addition, the degree of conversion through the thickness follows a gradient depending on the illumination time [45]. In this condition, the measurement of molecular absorption properties (ε, φ) is tricky. Nevertheless, it is still possible, by considering the local degree of conversion of the molecules inside the film or by using very thin films where the conversion can be considered uniform.

Supposing that the transparent form absorbs more than the colored one in the UV range of illumination, the UV penetration at the end of the conversion is determined by ε_C^{UV}, namely, the extinction coefficient of the colored form in the UV. In Figure 7, the case of a 10 μm thick film with a concentration of 400 mol/m^3 of molecule **6** is reported, showing the measured absorption spectra of the two forms and the calculated penetration depth as a function of the molar extinction coefficients.

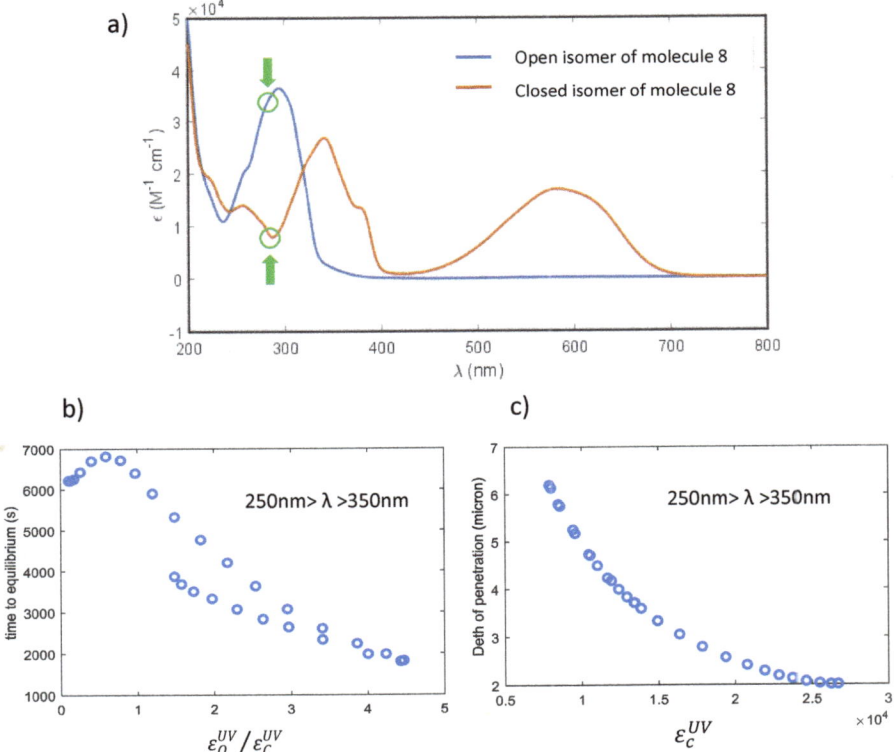

Figure 7. (a) UV-vis absorption spectra of the two isomers of molecule **6**; (b) time necessary to reach the photostationary state as function of the ratio $\varepsilon_O^{UV}/\varepsilon_C^{UV}$; (c) penetration depth as function of the irradiation wavelength. With the green circles the values of the molar extinction coefficients at the wavelength corresponding to the highest penetration are highlighted. Each point of figure (**b**) and (**c**) corresponds to a different simulation with a λ of irradiation changing between 250 nm and 350 nm.

The lower the absorption of the UV light by the colored form, the higher the penetration (Figure 7c). Fixed the quantum yield of the transformations (ϕ_{CO}^{UV}, ϕ_{OC}^{UV}), the time required to reach a stationary situation decreases while increasing the ratio $\varepsilon_O^{UV}/\varepsilon_C^{UV}$. It has been also demonstrated that this ratio affects the fatigue resistance of the diarylethenes [51].

All these considerations point out that the actual coloration of a photochromic material at the solid state depends not only on the intrinsic capability of absorbing visible light by the colored form, but also on its absorption at the illumination wavelength. This means that to reach large contrasts, large ε_C^{vis} cannot be the only selection criteria of a photochromic dye. If the absorption at the illumination wavelength (ε_C^{UV}) is high and comparable to the ε_C^{vis} (highlighted in Figure 8a for compound **13**), the penetration depth will be low, and the contrast will be similarly visible (C = 160). Instead, if the absorption is much lower, a consistent raise of the contrast value results (C = 3000 for compound **7**, Figure 8b).

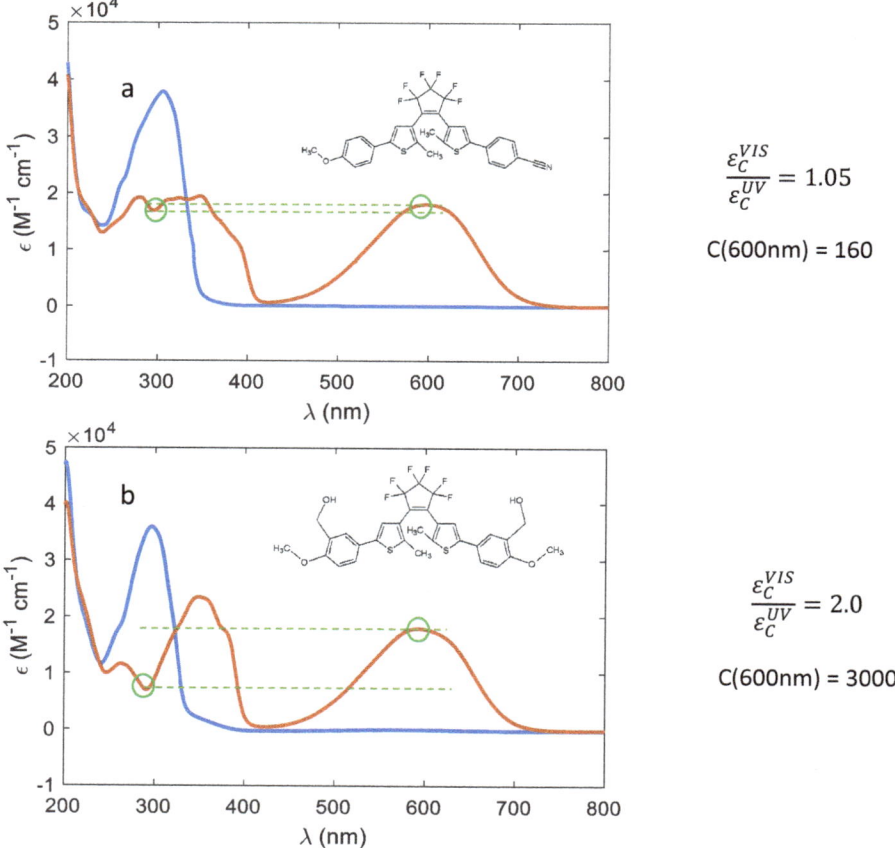

Figure 8. Comparison between two diarylethenes showing a different $\varepsilon_C^{vis}/\varepsilon_C^{UV}$ ratio: (**a**) Molecule **13**; (**b**) molecule **7**. On the left the molecules used in the simulations, together with their absorption properties are reported (the green circles highlight ε_C^{UV} and ε_C^{vis}); on the right, the value of the ratio $\varepsilon_C^{vis}/\varepsilon_C^{UV}$ and the contrast C computed at 600 nm are reported for a 10 μm film with a concentration of 300 mol/m^3.

3.2. Computational Tool

Once we had determined the relevant parameters that characterize the general photochromic behavior at the solid state, we made use of a kinetic model which describes both the coloration and the fading of a diarylethene film under specific illumination conditions [52], and we combined here all these pieces of information in a computational tool, which predicts its performance a priori. This allows for a proper selection of the photochromic material, which is necessary to satisfy the target properties of the optical elements (both phase and amplitude holograms), without a number of optimization experiments that would have been otherwise necessary, saving time and material.

In order to efficiently exploit the kinetic model, a Graphical User Interface (GUI) using Matlab® R2016b (The MathWorks, Natick, MA, USA) was built, facilitating the selection of the simulation parameters and displaying the desired results with a fast and practical routine. GUI is user-friendly, hence it can be used without any specific computational ability.

Two different versions of the tool were developed, enabling to simulate a photochromic film based on either one or two molecules, mixed together. In the following, the two molecules case is detailed, being the most complex one.

The program is organized in three different sections: (i) selection of the molecules, (ii) parameters choice and (iii) results visualization (Figure 9).

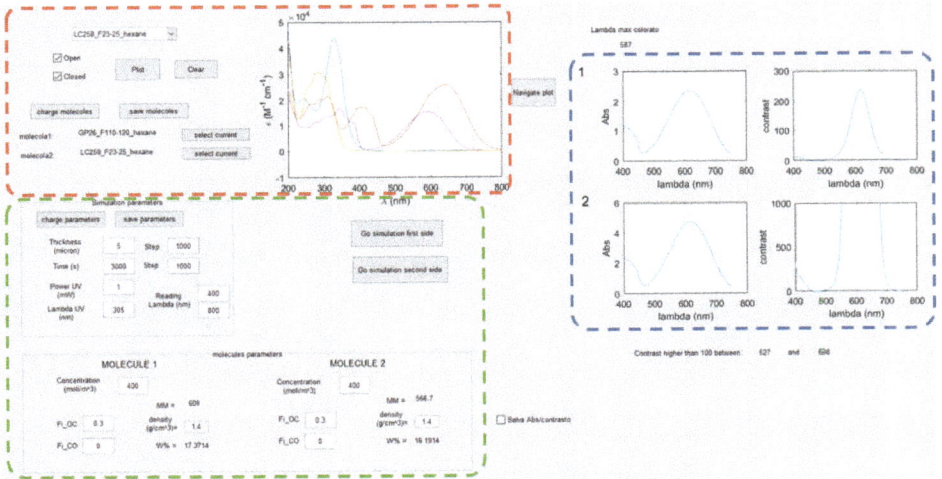

Figure 9. Screenshot of the GUI developed to simulate the absorption properties of photochromic films. The colored boxes indicate three different sections: selection of the molecules (red), parameters choice (green) and results visualization (blue).

3.2.1. Selection of the Molecules

The absorption properties in solution or solid state of the transparent and colored isomers (e.g., molar extinction coefficient as function of the wavelength) of a series of dyes are previously uploaded in a database. In this window, the UV-vis absorption spectra of one or two dyes picked up from this database are shown, allowing for an easy and rapid comparison (Figure 9 red box).

The key features usually considered are: (i) the wavelengths at which the photochromic transformation can be triggered; (ii) the correspondent absorbance values and the position and width of the absorbance band in the visible. Considered together, they give hints about the efficiency of the transformation and the possible final behavior of the materials.

3.2.2. Parameters Choice

In this window, the illumination conditions, including the wavelength and the illumination intensity, and the film thickness are set (Figure 9 green box). Setting the first parameters, the discretization of time and space is defined. In this section, we also set the wavelength range where the results will be computed.

On the material regard, the concentration of the selected molecule and its photochromic properties are selected, specifically the quantum yield of conversion at the illumination wavelength for the direct (ϕ_{OC}) and inverse (ϕ_{CO}) transformation, and the density of the material (usually considered equal to the density of the chosen polymer matrix). In the case of two dyes, the concentration of molecule 1 and molecule 2 is set.

3.2.3. Visualization of the Results

Once we had selected the molecules and the simulation parameters, the computation could start. The program is divided into two steps, enabling us to simulate a double illumination process, with one process for each side of the film. At each stage, the concentration profile of the transparent isomer is plotted, as function of time and space, and the limit of the UV penetration is computed. By the analysis of these plots, it is possible to understand if either the molecular concentrations or the film thickness is too large to have a complete conversion inside the sample volume. In Figure 10, we show examples where the total conversion inside the film volume is reached or not, depending on the material thickness. The plots show the transparent isomer concentration in dependence of time and thickness, after one (a1, b1) and two (a2, b2) sides were illuminated.

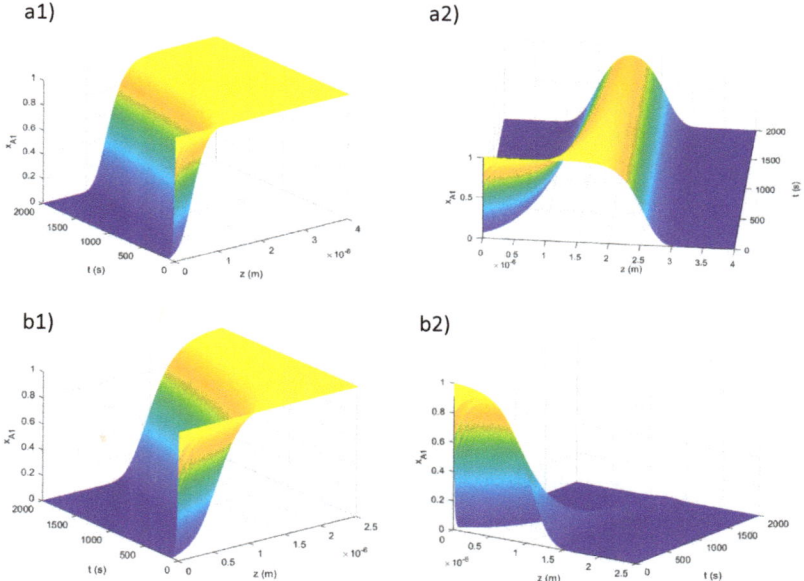

Figure 10. Concentration profile of the transparent isomer after the simulated irradiation of one (**a1,b1**) and two sides (**a2,b2**). The plots show the time evolution of the profile and the penetration inside the volume. In this example, we show a film of 4 μm, where it is impossible to convert all the material (**a2**), but lowering the value to 2.5 μm means the total conversion is reached after the two side illumination (**b2**). Photostationary state after 2000 s.

In the case reported in Figure 10a1, we notice that a large amount of the film is not converted by a single side illumination. Also with a double side illumination (Figure 10a2), the inner part of the film remains unconverted. In the case b, a single side exposure is not enough to achieve the total conversion, but a double illumination (Figure 10b2) induces a complete coloration.

In Figure 11, we report an example of how the absorbance properties of two different dyes can be combined to reach higher contrast performances.

In this case, two molecules with a visible absorbance around 550 nm and 650 nm are combined to cover a wide wavelength range. In panel b), we report the absorbance and relative contrast of the film, after one and two sides exposure considering a thickness of 4 µm and a concentration of about 16 wt % for both dyes.

We notice that the absorbance almost double going from one side to the both sides illumination, meaning that the penetration depth is roughly half of the film thickness. As for the contrast (plots on the right), the increase is very large by the double exposure, reaching values larger than 1000 in a range wider than 100 nm. We also notice that in the 400 nm region, the contrast is quite good (>100) thanks to the presence of the secondary peak in mol1 (Figure 11a) and only a small spectral region around 450 nm has a low absorbance.

3.3. Remarks on the Absorption Properties

According to the discussions we reported, it is clear how useful such tool can be in designing high performance photochromic films and how many pieces of information can be retrieved from the simulations. Given a desired contrast in a certain wavelength range for a specific application, this tool supports the choice of the right set of molecules to be used. Once they have been selected, the illumination wavelength has to be carefully chosen: ε_C^{UV} should be low, to have a deep penetration through the film, thus achieving a full conversion; moreover, the $\varepsilon_O^{UV}/\varepsilon_C^{UV}$ and $\varepsilon_C^{vis}/\varepsilon_C^{UV}$ ratios should be high in order to have a fast kinetic and high contrast.

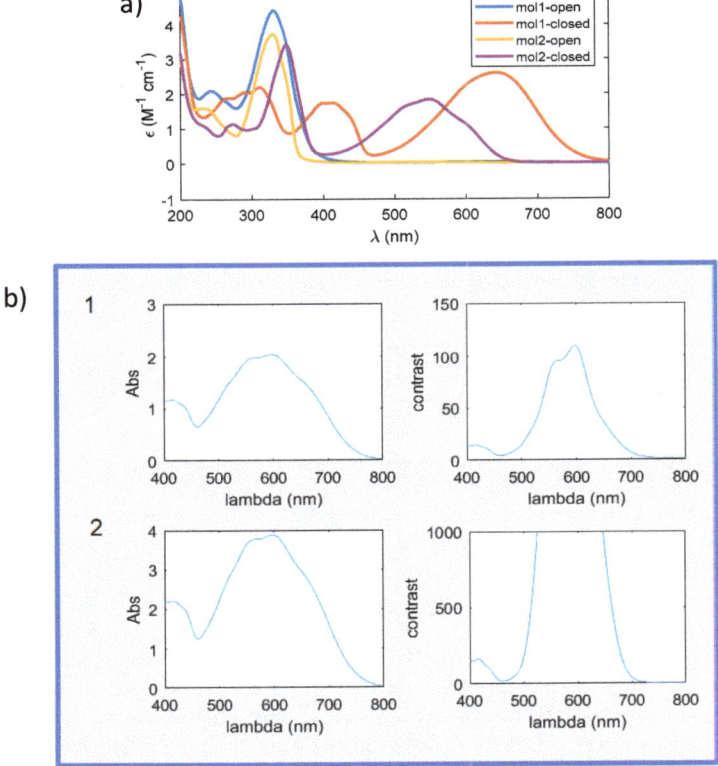

Figure 11. (a) Molar extinction coefficient of two diarylethenes (molecule **12** and **24**) used for the simulation; (b) Screenshot of the results computed: the figure shows absorbance and contrast of the film after one and two exposures (performed on two sides) in the spectral range of interest.

In the case of a mix of two dyes, the choice must be particularly careful. Both molecules have to be efficiently converted at the same time with the same illumination wavelength, meaning that they should have similar absorbing profiles in the UV region. Otherwise, one dye could behave as a barrier for the conversion of the other one, resulting in lower and unexpected absorption performances.

Finally, information on the concentration of the dye and the film thickness are provided to fulfill the requirement of contrast. With the proper selection of the diarylethene and films of few microns of a polymer matrix containing 20–25% of chromophore or backbone photochromic polymers, it is possible to reach a suitable contrast for the application herein reported.

3.4. Refractive Index Modulation

In the case of volume phase devices, we showed that a modulation of the refractive index is necessary in order to induce a controlled phase delay, which is equal to the product refractive index n times the film thickness d. Considering a target product $d \cdot \Delta n$ of about 0.4–1 µm, values of Δn = 0.04–0.1 are required for film thicknesses in the range of 1–10 µm.

In order to understand this requirement from a materials point of view, we start from the Lorentz-Lorenz equation. This equation [53] links the macroscopic refractive index with the material density and the molecular polarizability α (since we are in the optical spectral region only electronic polarizability is considered):

$$N = \frac{n^2-1}{n^2+2} = \frac{4\pi}{3} \frac{N_A}{V} \alpha \tag{7}$$

where n is the material refractive index, N_A the Avogadro number, V is the molar volume. This is valid for a monocomponent material, but photochromic films are multicomponent and the refractive index contains a contribution of both the matrix and the photochromic dye. Considering no interaction between these two components, we can write an effective refractive index as the sum of their contributions:

$$\frac{n^2-1}{n^2+2} = C_{matrix}\frac{n^2_{matrix}-1}{n^2_{matrix}+2} + C_{dye}\frac{n^2_{dye}-1}{n^2_{dye}+2} = C_{matrix}N_{matrix} + C_{dye}N_{dye} \tag{8}$$

$$C_{matrix} + C_{dye} = 1 \tag{9}$$

where n_{dye}, n_{matrix} are the refractive indices of a material composed by the pure dye and the polymer matrix respectively and C_{matrix}, C_{dye} are the relative volume concentrations. Accordingly, the refractive index of the colored (or uncolored) material is then:

$$n_{c(o)} = \sqrt{\frac{2C_{dye}\left(N^{c(o)}_{dye} - N_{matrix}\right) + 2N_{matrix} + 1}{1 - N_{matrix} - C_{dye}\left(N^{c(o)}_{dye} - N_{matrix}\right)}} \tag{10}$$

We noticed that the refractive index of the film depends on the contrast between the value of the matrix and of the photochromic dye. Usually, the matrix shows a refractive index lower than the value of the photochromic dye. Even more important for determining the refractive index of the doped film is the concentration of the photochromic species. It must be as large as possible, but avoiding any side effects such as segregation or aggregation. Actually, we are interested in the change in the refractive index going from one photochromic form to the other. Looking again to the Equation (7), we notice that a large change in the molecular polarizability α between the two forms is necessary, in addition to the previous requirements, to enhance the modulation in the refractive index. The polarizability is proportional to the number of electrons in the molecule, but it is known that π conjugated systems exhibit higher polarizability, and the enhancement is proportional to the degree of delocalization [54]. Considering diarylethenes, it is apparent that the closed (colored) form is more conjugated than the open (uncolored) form; therefore, it shows a larger refractive index. The presence of electroactive substituents can play a role in increasing the modulation of the refractive index [55]. Moreover, the molecular polarizability is wavelength dependent as the refractive index. In the optical regime, it increases with the frequency in a marked way approaching the resonance frequencies due to the electronic transitions [56]. Since the colored form shows visible absorption bands, this pre-resonance effect will be more important than for the uncolored form (only UV absorptions). Moreover, a steeper increase in the molecular polarizability takes place in the NIR. Consequently, the modulation of the refractive index will benefit from this effect and it increases with the frequency too [57]. This feature has been recently highlighted [49] in a series of diarylethene based polyurethanes, where a clear positive trend existed between the Δn and the absorption wavelength of the colored form.

To sum up, in order to maximize the modulation, it will be necessary to:

- maximize the concentration of the photochromic dye in the film;
- design a photochromic molecule with specific chemical groups that enhance the change in the molecular polarizability (large change in the π conjugation path and efficiency);
- increase the wavelength gap between the absorption band in the UV of the uncolored form and the visible band of the colored form.

According to the experimental results reported in the literature, the concentration parameter is the most important one in affecting the Δn and for these reasons, backbone photochromic polymers have been developed. Values of the order of 0.08 at 800 nm were measured [58].

4. Writing Strategies and Examples of CGHs

In a typical route for the CGH production, the photochromic film is converted to the colored form by irradiation with UV light. Then, the layer is patterned upon exposure to visible light, which induces a selective bleaching of the film. We considered two different strategies here for the substrate patterning: (i) a mask projection system, based onto a spatial light modulator; (ii) a scanning system, by direct laser writing (maskless lithography). The two techniques, presented hereafter, are complementary. In both cases, the writing process may not introduce imperfections, called pattern distortions, especially when the holograms are used in interferometric applications. They are basically due to a misalignment of the writing beam with respect to its ideal position and can be quantified as the introduced wavefront error ΔW_ζ [59]:

$$\Delta W_\zeta = -\frac{m\lambda\zeta}{G} \qquad (11)$$

where m is the diffraction order, ζ is the grating position error in the direction perpendicular to the pattern lines and G is the local line spacing. To minimize these errors, it is convenient to work at low diffraction orders and with coarse line patterns. Along with this, the quality of the reconstructed image depends on the planarity of the substrate, since any imperfection produces phase contaminations. This is valid for the substrate itself, as well as for the photochromic film. It is crucial, accordingly, to optimize the depositing process not to introduce high spatial frequency errors in the transmitted wavefront, for both the film thickness and planarity.

4.1. Mask Projection

This approach consists in the projection of a mask specifically designed with the target pattern onto the photochromic substrate. Such approach derives directly from the well-established mask lithography [60]. An interesting possibility consists of the image projection through an Offner relay, which produces a one to one projection of the mask plane, where a Digital Micromirror Device (DMD) is placed, onto the sample plane, where the photochromic film is [61]. A DMD is a rectangular pattern of micromirrors that can be independently addressed between two specific angular positions. The device used in our tests by Texas Instruments (Dallas, TX, USA), is composed by 2048 × 1080 micro-mirrors with a pitch of 13.64 μm. The optical quality of the system is limited by the micromirror size and not by the optical aberrations. During the writing process, the DMD is homogeneously illuminated by a filtered light source. A CGH imaging system is also present, to follow the writing process in real time (Figure 12).

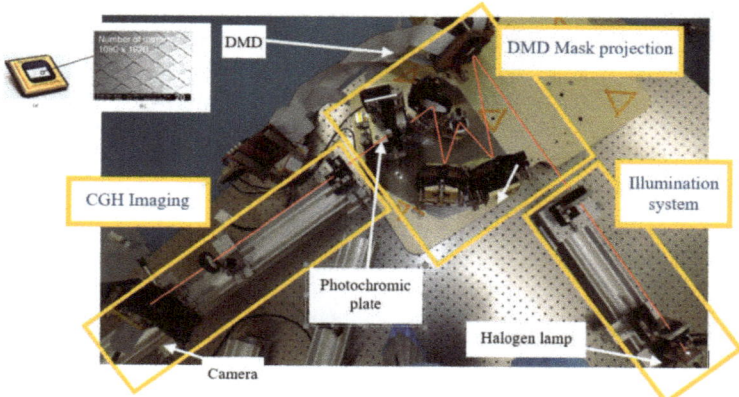

Figure 12. DMD based set-up for writing the photochromic CGHs. The three main subsystems, namely illuminating system, DMD mask projection and CGH imaging are highlighted. In the inset, a picture of the DMD used is reported [61].

Recording a binary CGH requires the projection of a single DMD mask for enough time to produce the full conversion of the film from the opaque to the transparent form. The advantage of the DMD projection system is the possibility to easily write grayscale CGHs [62]. Since the DMD is a programmable device, any mask can be projected for a specific amount of time. In fact, the photochromic material becomes progressively transparent when illuminated by visible light, and a given level of transparency, i.e., a given level of gray, is obtained with a well-defined exposure time. Caution is needed since the material response is not linear with the exposure time, but the transmission curve as function of the expose time can be measured before the CGH production and used for the linearization. Figure 13 illustrates an example of a grayscale CGH, with four different masks used for its realization.

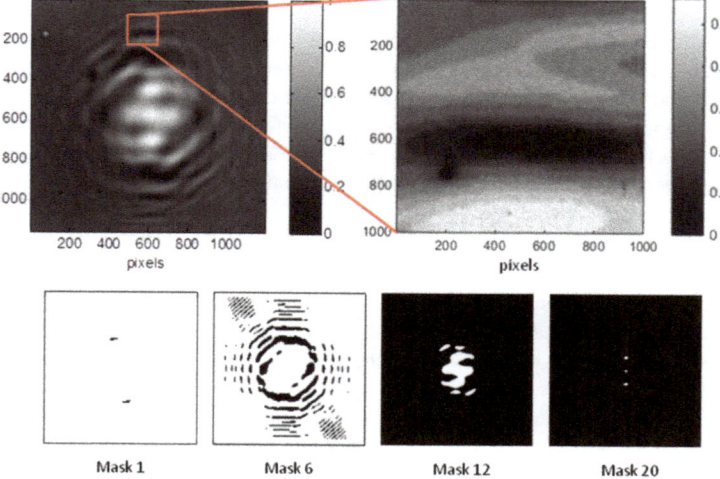

Figure 13. A grayscale CGH, the magnification shows the well-defined gray levels; examples of masks used for its production [62].

The versatility of the DMD is the great advantage of this technique: although a binary amplitude CGH has a higher efficiency, a grayscale hologram enables for much better image reconstruction quality, which leads to a better control on the wavefront generated by the CGH. On the other hand, the image resolution is limited by the micromirror size and number, i.e., by the ratio between the dimensions of the single micromirror and the whole chip. A possible step forward could be the stitching of different DMD projections to create a larger CGH or to demagnify the DMD, thus increasing the resolution (the image of the single mirror is smaller) and then stitching the different DMD images.

4.2. Direct Laser Writing

With direct laser writing, the pattern is transferred to the photosensitive layer using a light beam focalized in a theoretically diffraction limited spot onto the substrate. The laser power can be continuously adjusted while the substrate is scanned in the plane and exposed where necessary. Usually, an autofocus system keeps the substrate in the correct axial position to guarantee the best spot resolution. In the past, we investigated the possibility to use commercial direct laser machines to transfer patterns onto photochromic substrates [63], but we faced problems due to the writing speed, light power, and wavelength. In fact, commercial systems are characterized by high speed rates (hundreds of mm/s), very high light powers (tens of mW/μm^2), and usually work in the spectral region suitable for photoresists, namely in the UV, which is not really suitable for diarylethenes. We observed a low definition of the pattern, and the formation of surface reliefs on the coating given by the local heating of the substrate. In contrast, the resolution was very high, being limited by the spot size (down to 1 μm). We therefore developed custom direct laser machines for the production of photochromic CGHs, where we optimized the writing speed, the light power and the writing wavelength.

The developed system is shown in Figure 14 [64]. It is composed by a moving table (raster X-Y scan) and an optical bench (Offner relay layout), mounted vertically on a fixed bridge. The light source is a multichannel laser system equipped with four heads at 406 nm, 520 nm, 638 nm and 685 nm. The different wavelengths were selected as function of the sensitivity curve of photochromic materials and can be used independently. The light is coupled to an optical fiber and guided to the writing head. A trigger mechanism driven by the linear stage switches the lasers on and off at MHz speed. A viewing camera is also present to align the substrate and follow the writing process. The spot size is 3–4 μm depending on the wavelength, the writing speed 1–3 mm/s, and the laser power at the focal plane 1–3 mW.

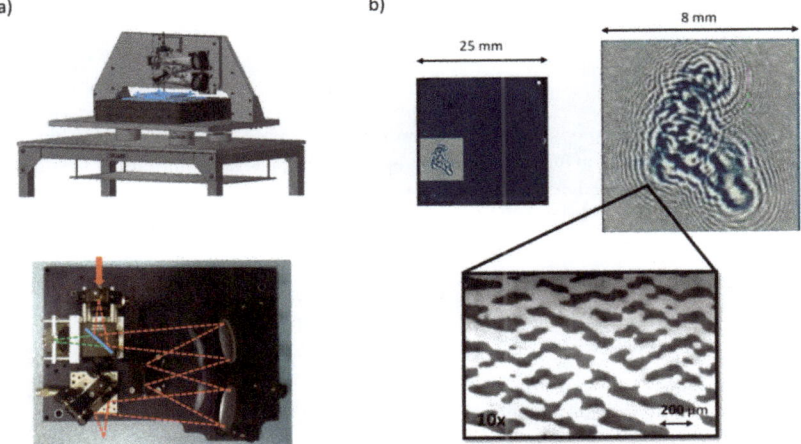

Figure 14. (a) Mechanical (top) and optical (bottom) schemes of our direct laser writing machine; (b) Different magnifications of a typical sample of photochromic film, written with a Fresnel CGH [64].

4.3. Examples of Diarylethene-based CGHs

Here we show some examples of holograms obtained with photochromic films based on diaryletehenes. As previously discussed, the calculated hologram phase function can be approximated as an amplitude or phase pattern, both binary and grayscale, and transferred to the photochromic layer with the most appropriate technique. While for binary CGHs direct laser writing is preferred, grayscale CGHs can be more easily obtained with the mask projection technique. In the latter case, recording a binary CGH requires the projection of a single mask to the photochromic plate, while grayscale holograms can be obtained by sequentially displaying a series of binary masks to locally create the desired level of transparency [61]. Considering diarylethenes, amplitude holograms performs well in the visible region, approximately between 500 and 800 nm, while phase holograms performed well in the NIR region, approximately between 800 and 1500 nm.

A nice example is the CGH of a Fresnel lens reported in Figure 15 [49]. This CGH behaves as a spherical lens and the focal length is dependent on the spacing of the lines. As clearly shown in Figure 15b, it is possible to identify a focused spot on the camera both illuminating the CGH with a red laser (650 nm), where the hologram behaves as an amplitude hologram, and illuminating the CGH with a NIR laser (980 nm), where the hologram is a pure phase hologram. The corresponding transmission spectra and the refractive index dispersion curves are reported in Figure 15a, where it is marked with arrows showing the change of property between the two forms in the film.

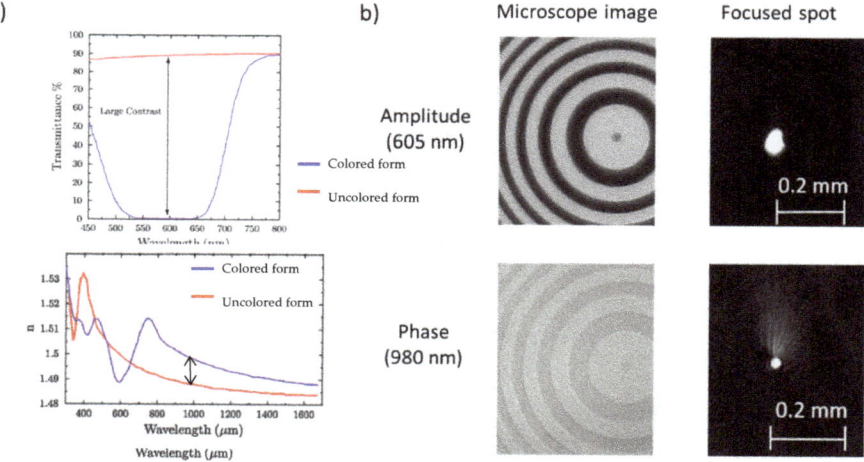

Figure 15. (**a**) Measured modulation of transmittance and refractive index of a high content photochromic film; transparent form (red line) and opaque form (blue line). The arrows highlight the change in the transparency and refractive index; (**b**) Left: Microscope images (phase and amplitude) of a Fresnel CGH recorded on a diarylethene based film. Right: CCD images of the laser spot in the focal plane [49].

Another example of photochromic CGHs is reported in Figure 16. It is the image of a dandelion (430 × 430 pixels), that has been transferred by direct laser writing. The hologram is a binary amplitude type, square with a 17 mm side, and a pixel size of 3 μm.

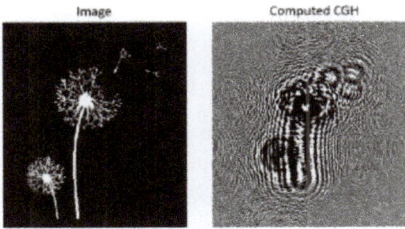

Figure 16. Image of a dandelion and the calculated binary Fresnel CGH (4 × 4 mm² size at a focus of 0.5 m) [64].

The calculated CGH was transferred on the photochromic film by means of the direct laser writing machine and the result is reported in Figure 17 details of the pattern.

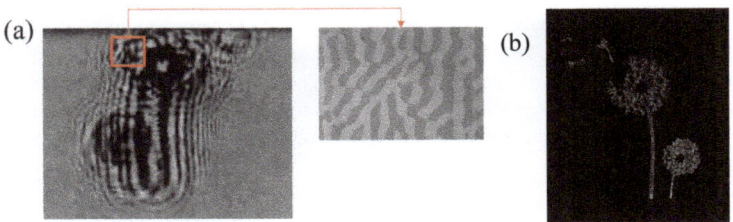

Figure 17. (**a**) Photograph of the CGH and a magnification of the written pattern; (**b**) The reconstructed image at 633 nm [64].

The image is reconstructed at 0.5 m, with a size of 4 × 4 mm² (Figure 17b). We notice the complexity of the image with small details, which requires a large hologram due to the high information density on its edge. Hologram resolution and size prevented the use of the mask projection technique to obtain the same level of details.

The second example is the image of a letter "Z" (200 × 200 pixels), obtained by mask projection. Accordingly, the hologram is grayscale amplitude. The size of the CGH is limited to 10 × 10 mm², which leads to a CGH resolution of 720 × 720 pixels according to the DMD size. In order to be sure that all the fringes in the CGHs are resolved, the image physical size and the focus are fixed at 2 × 2 mm² and 2 m, respectively. Once we obtained the continuous complex pattern, its magnitude was discretized to twenty gray levels with thresholds ranging from 0 to 1 in steps of 0.05.

Figure 18 shows the calculated and the actual grayscale CGH of the letter Z, along with the theoretical and experimental reconstructed image.

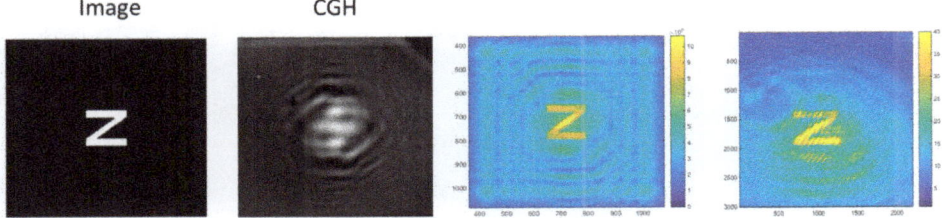

Figure 18. From left to right: the calculated and recorded (grayscale) CGH on the photochromic polymer; simulated and experimentally reconstructed image at 633 nm [62].

The image, reconstructed at 633 nm, shows a dimension on the camera of 2 × 2 mm², as expected. Very faithful reconstruction has been obtained with respect to the simulated reconstructed image as well as the original "Z" image. Also in this case, we can notice the fidelity of the reconstruction, confirming the effectiveness of the mask projection approach to produce amplitude grayscale holograms.

5. Conclusions

Photochromic materials give interesting opportunities as substrates for the manufacturing of rewritable Computer-Generated Holograms (CGHs). Indeed, phase and amplitude holograms are demonstrated and binary and grayscale pattern can be easily transferred. In order to make high quality holograms, the optimization of both the photochromic material and the writing procedure is necessary. As for the material optimization, the combination of a kinetic model and experimental UV-vis data makes possible the development of a computational tool to predict the performances in terms of transparency contrast of photochromic films. In this way, a balanced choice of the film thickness and photochromic content leads to high efficiency amplitude holograms. In addition, the versatility in the synthesis of photochromic diarylethenes provides many possibilities in tuning the spectral position and intensity of the band along the whole visible region. As for the phase hologram and the modulation of the refractive index, important guidelines are provided in order to maximize the efficiency. We also highlighted that another crucial aspect is the writing strategy; here both a reconfigurable mask approach based on a DMD chip and a direct laser writing machine are reported. The former is more suitable for the realization of grayscale patterns, but suffers from a low spatial resolution; the latter is more suitable for binary patterns and provides a much larger spatial resolution in the case of large area CGHs.

Author Contributions: Introduction, writing—original draft preparation, supervision A.B.; writing—review and editing, all authors; modelling and software L.O., G.P.; materials analysis, C.B.; experimental work and data analysis, L.O., G.P., F.Z.

Funding: This research received no external funding.

Acknowledgments: The authors would like to thank Mariachiara Mantero for the support in direct writing hologram realization.

Conflicts of Interest: The authors declare no conflict of interest.

References

1. Gabor, D. A new microscopic principle. *Nature* **1948**, *161*, 777–778. [CrossRef] [PubMed]
2. Bertolotti, M. *The History of the Laser*; CRC Press: Boca Raton, FL, USA, 2004.
3. Johnston, S.F. Reconstructing the history of holography. *Proc. SPIE* **2003**, *5005*, 455–464.
4. Lawrence, J.R.; Neill, F.T.O.; Sheridan, J.T. Photopolymer holographic recording material. *Optik* **2001**, *112*, 449–463. [CrossRef]
5. Bianco, A.; Perissinotto, S.; Garbugli, M.; Lanzani, G.; Bertarelli, C. Control of optical properties through photochromism: A promising approach to photonics. *Laser Photonics Rev.* **2011**, *5*, 711–736. [CrossRef]
6. Kobayashi, Y.; Abe, J. Real-Time Dynamic Hologram of a 3D Object with Fast Photochromic Molecules. *Adv. Opt. Mater.* **2016**, *4*, 1354–1357. [CrossRef]
7. Boiko, Y. Volume hologram recording in diarylethene doped polymer. *Opt. Mem. Neural Netw.* **2008**, *17*, 30–36.
8. Pu, S.; Yang, T.; Yao, B.; Wang, Y.; Lei, M.; Xu, J. Photochromic diarylethene for polarization holographic optical recording. *Mater. Lett.* **2007**, *61*, 855–859. [CrossRef]
9. Yan, P.; Tong, Z.; Pu, S.; Liu, W. A new unsymmetrical diarylethene for reversible holographic recording. *Adv. Manuf. Technol.* **2011**, *157*, 655–659. [CrossRef]
10. Luo, S.; Chen, K.; Cao, L.; Liu, G.; He, Q.; Jin, G.; Zeng, D.; Chen, Y. Photochromic diarylethene for rewritable holographic data storage. *Opt. Express* **2005**, *13*, 3123–3128. [CrossRef]
11. Cao, L.; Wang, Z.; Zong, S.; Zhang, S.; Zhang, F.; Jin, G. Volume Holographic Polymer of Photochromic Diarylethene for Updatable Three-Dimensional Display. *J. Polym. Sci. Part B Polym. Phys.* **2016**, *54*, 2050–2058. [CrossRef]

12. Kajimura, M.; Egami, C. Three-dimensional multiplex micro-hologram using diarylethene-doped PMMA film. *Mol. Cryst. Liq. Cryst.* **2016**, *635*, 102–106. [CrossRef]
13. Irie, M. Photochromism of diarylethene molecules and crystals. *Proc. Jpn. Acad. Ser. B* **2010**, *86*, 472–483. [CrossRef] [PubMed]
14. Nishi, H.; Kobatake, S. Photochromism and Optical Property of Gold Nanoparticles Covered with Low-Polydispersity Diarylethene Polymers. *Macromolecules* **2008**, *41*, 3995–4002. [CrossRef]
15. Nishi, H.; Asahi, T.; Kobatake, S. Light-Controllable Surface Plasmon Resonance Absorption of Gold Nanoparticles Covered with Photochromic Diarylethene Polymers. *J. Phys. Chem. C* **2009**, *113*, 17359–17366. [CrossRef]
16. Tsuboi, Y.; Shimizu, R.; Shoji, T.; Kitamura, N. Near-Infrared Continuous-Wave Light Driving a Two-Photon Photochromic Reaction with the Assistance of Localized Surface Plasmon. *J. Am. Chem. Soc.* **2009**, *131*, 12623–12627. [CrossRef]
17. Piard, J.; Métivier, R.; Giraud, M.; Léaustic, A.; Yu, P.; Nakatani, K. Photoswitching in diarylethene nanoparticles, a trade-off between bulk solid and solution: Towards balanced photochromic and fluorescent properties. *New J. Chem.* **2009**, *33*, 1420–1426. [CrossRef]
18. Spangenberg, A.; Métivier, R.; Yasukuni, R.; Shibata, K.; Brosseau, A.; Grand, J.; Aubard, J.; Yu, P.; Asahi, T.; Nakatani, K. Photoswitchable interactions between photochromic organic diarylethene and surface plasmon resonance of gold nanoparticles in hybrid thin films. *Phys. Chem. Chem. Phys.* **2013**, *15*, 9670–9678. [CrossRef]
19. Snegir, S.V.; Khodko, A.A.; Sysoiev, D.; Lacaze, E.; Pluchery, O.; Huhn, T. Optical properties of gold nanoparticles decorated with furan-based diarylethene photochromic molecules. *J. Photochem. Photobiol. A Chem.* **2017**, *342*, 78–84. [CrossRef]
20. Tsutsumi, N. Molecular design of photorefractive polymers. *Polym. J.* **2016**, *48*, 571–588. [CrossRef]
21. Lynn, B.; Blanche, P.; Peyghambarian, N. Photorefractive Polymers for Holography. *J. Polym. Sci. Part B Polym. Phys.* **2014**, *52*, 193–231. [CrossRef]
22. Marinova, V.; Huei Lin, S.; Chung Liu, R.; Hsu, K.Y. Photorefractive Effect: Principles, Materials, and Near-Infrared Holography. In *Wiley Encyclopedia of Electrical and Electronics Engineering*; John Wiley & Sons, Inc.: Hoboken, NJ, USA, 2016; pp. 1–20.
23. Blanche, P.-A.; Bigler, C.; Ka, J.-W.; Peyghambarian, N. Fast and continuous recording of refreshable holographic stereograms. *Opt. Eng.* **2018**, *57*, 061608. [CrossRef]
24. Blanche, P.-A.; Bablumian, A.; Voorakaranam, R.; Christenson, C.; Lin, W.; Gu, T.; Flores, D.; Wang, P.; Kathaperumal, M.; Rachwal, B.; et al. Holographic three-dimensional telepresence using large-area photorefractive polymer. *Nature* **2010**, *468*, 80–83. [CrossRef] [PubMed]
25. Tay, S.; Blanche, P.A.; Voorakaranam, R.; Tunc, A.V.; Lin, W.; Rokutanda, S.; Gu, T.; Flores, D.; Wang, P.; Li, G.; et al. An updatable holographic three-dimensional display. *Nature* **2008**, *451*, 694–698. [CrossRef] [PubMed]
26. Hesselink, L.; Orlov, S.S.; Liu, A.; Akella, A.; Lande, D.; Neurgaonkar, R.R. Photorefractive materials for nonvolatile volume holographic data storage. *Science* **1998**, *282*, 1089–1094. [CrossRef] [PubMed]
27. Shishido, A. Rewritable holograms based on azobenzene-containing liquid-crystalline polymers. *Polym. J.* **2010**, *42*, 525–533. [CrossRef]
28. Ryabchun, A.; Bobrovsky, A. Cholesteric Liquid Crystal Materials for Tunable Diffractive Optics. *Adv. Opt. Mater.* **2018**, *6*, 1800335. [CrossRef]
29. Matharu, A.S.; Jeeva, S.; Ramanujam, P.S. Liquid crystals for holographic optical data storage. *Chem. Soc. Rev.* **2007**, *36*, 1868–1880. [CrossRef] [PubMed]
30. Liu, Y.J.; Sun, X.W. Holographic Polymer-Dispersed Liquid Crystals: Materials, Formation, and Applications. *Adv. Optoelectron.* **2008**, *2008*, 684349. [CrossRef]
31. De Sio, L.; Lloyd, P.F.; Tabiryan, N.V.; Bunning, T.J. Hidden Gratings in Holographic Liquid Crystal Polymer-Dispersed Liquid Crystal Films. *ACS Appl. Mater. Interfaces* **2018**, *10*, 13107–13112. [CrossRef]
32. Brown, B.R.; Lohmann, A.W. Computer-generated Binary Holograms. *IBM J. Res. Dev.* **1969**, *13*, 160–168. [CrossRef]
33. Tricoles, G. Computer generated holograms: An historical review. *Appl. Opt.* **1987**, *26*, 4351–4360. [CrossRef] [PubMed]
34. Jacubowiez, L. Back to basics: Computer-generated holograms. *Photoniques* **2017**, *2*, 38–44. [CrossRef]
35. Bass, M. *Handbook of Optics: Volume II*; McGraw-Hill: New York, NY, USA, 2010.

36. Ruffato, G.; Rossi, R.; Massari, M.; Mafakheri, E.; Capaldo, P.; Romanato, F. Design, fabrication and characterization of Computer Generated Holograms for anti-counterfeiting applications using OAM beams as light decoders. *Sci. Rep.* **2017**, *7*, 18011. [CrossRef] [PubMed]
37. MacGovern, A.J.; Wyant, J.C. Computer Generated Holograms for Testing Optical Elements. *Appl. Opt.* **1971**, *10*, 619–624. [CrossRef] [PubMed]
38. Fang, F.Z.; Zhang, X.D.; Weckenmann, A.; Zhang, G.X.; Evans, C. Manufacturing and measurement of freeform optics. *CIRP Ann. Manuf. Technol.* **2013**, *62*, 823–846. [CrossRef]
39. Kang, G.G.; Xie, J.; Liu, Y. New design techniques and alignment methods for CGH-null testing of aspheric surface. *Proc. SPIE* **2008**, *6624*, 66240K.
40. Li, S.; Zhang, J.; Liu, W.; Guo, Z.; Li, H.; Yang, Z.; Liu, B.; Tian, A.; Li, X. Measurement investigation of an off-axis aspheric surface via a hybrid compensation method. *Appl. Opt.* **2018**, *57*, 8220–8227. [CrossRef]
41. Poon, T.-C.; Liu, J.-P. *Introduction to Modern Digital Holography with Matlab*; Cambridge University Press: Cambridge, UK, 2014.
42. Lee, W.H. Sampled Fourier Transform Hologram Generated by Computer. *Appl. Opt.* **1970**, *9*, 639–643. [CrossRef]
43. Zamkotsian, F.; Alata, R.; Lanzoni, P.; Pariani, G.; Bianco, A.; Bertarelli, C. Programmable CGH on photochromic material using DMD generated masks. *Proc. SPIE* **2018**, *10546*, 1054606.
44. Neto, L.G.; Cardona, P.S.P.; Cirino, G.A.; Mansano, R.D.; Verdonck, P.B. Implementation of Fresnel full complex-amplitude digital holograms. *Opt. Eng.* **2004**, *43*, 2640–2649. [CrossRef]
45. Bianco, A.; Pariani, G.; Zanutta, A.; Castagna, R.; Bertarelli, C. Photochromic materials for holography: Issues and constraints. *Proc. SPIE* **2012**, *8281*, 828104.
46. Baldry, I.K.; Bland-Hawthorn, J.; Robertson, J.G. Volume phase holographic gratings: Polarization properties and diffraction efficiency. *Publ. Astron. Soc. Pac.* **2004**, *116*, 403–414. [CrossRef]
47. Irie, M.; Mohri, M. Thermally Irreversible Photochromic Systems. Reversible Photocyclization of Diarylethene Derivatives. *J. Org. Chem.* **1988**, *53*, 803–808. [CrossRef]
48. Irie, M.; Fukaminato, T.; Matsuda, K.; Kobatake, S. Photochromism of diarylethene molecules and crystals: Memories, switches, and actuators. *Chem. Rev.* **2014**, *114*, 12174–12277. [CrossRef] [PubMed]
49. Oggioni, L.; Toccafondi, C.; Pariani, G.; Colella, L.; Canepa, M.; Bertarelli, C.; Bianco, A. Photochromic Polyurethanes Showing a Strong Change of Transparency and Refractive Index. *Polymer* **2017**, *9*, 462. [CrossRef] [PubMed]
50. Suzuki, H. *Electronic Absorption Spectra and Geometry of Organic Molecules: An Application of Molecular Orbital Theory*; Academic Press Inc.: Cambridge, MA, USA, 1967.
51. Pariani, G.; Quintavalla, M.; Colella, L.; Oggioni, L.; Castagna, R.; Ortica, F.; Bertarelli, C.; Bianco, A. New Insight into the Fatigue Resistance of Photochromic 1,2-Diarylethenes. *J. Phys. Chem. C* **2017**, *121*, 23592–23598. [CrossRef]
52. Pariani, G.; Bianco, A.; Castagna, R.; Bertarelli, C. Kinetics of photochromic conversion at the solid state: quantum yield of dithienylethene-based films. *J. Phys. Chem. A* **2011**, *115*, 12184–12193. [CrossRef] [PubMed]
53. Bottcher, C.J.F. *Theory of Electric Polarisation*; Elsevier: Amsterdam, The Netherland, 1952.
54. Gussoni, M.; Rui, M.; Zerbi, G. Electronic and relaxation contribution to linear molecular polarizability. An analysis of the experimental values. *J. Mol. Struct.* **1998**, *447*, 163–215. [CrossRef]
55. Bertarelli, C.; Bianco, A.; D'Amore, F.; Gallazzi, M.C.; Zerbi, G. Effect of Substitution on the Change of Refractive Index in Dithienylethenes: An Ellipsometric Study. *Adv. Funct. Mater.* **2004**, *14*, 357–363. [CrossRef]
56. Djorovic, A.; Meyer, M.; Darby, B.L.; Ru, E.C. Le Accurate Modeling of the Polarizability of Dyes for Electromagnetic Calculations. *ACS Omega* **2017**, *2*, 1804–1811. [CrossRef]
57. Bertarelli, C.; Bianco, A.; Castagna, R.; Pariani, G. Journal of Photochemistry and Photobiology C: Photochemistry Reviews Photochromism into optics: Opportunities to develop light-triggered optical elements. *J. Photochem. Photobiol. C Photochem. Rev.* **2011**, *12*, 106–125. [CrossRef]
58. Toccafondi, C.; Occhi, L.; Cavalleri, O.; Penco, A.; Castagna, R.; Bianco, A.; Bertarelli, C.; Comoretto, D.; Canepa, M. Photochromic and photomechanical responses of an amorphous diarylethene-based polymer: A spectroscopic ellipsometry investigation of ultrathin films. *J. Mater. Chem. C* **2014**, *2*, 4692–4698. [CrossRef]
59. Chang, Y.R.; Burge, J.H. Error analysis for CGH optical testing. *Proc. SPIE* **1999**, *3782*, 358–366.
60. Lin, B.J. *Optical Lithography: Here Is Why*; SPIE: Bellingham, DC, USA, 2010.

61. Alata, R.; Pariani, G.; Zamkotsian, F.; Lanzoni, P.; Bianco, A.; Bertarelli, C. Programmable CGH on photochromic material using DMD. *Proc. SPIE* **2016**, *9912*, 991234.
62. Alata, R.; Pariani, G.; Zamkotsian, F.; Lanzoni, P.; Bianco, A.; Bertarelli, C. Programmable CGH on photochromic plates coded with DMD generated masks. *Opt. Express* **2017**, *25*, 6945–6953. [CrossRef]
63. Pariani, G.; Bertarelli, C.; Bianco, A.; Schaal, F.; Pruss, C. Characterization of photochromic Computer Generated Holograms for optical testing. *Proc. SPIE* **2012**, *8450*, 845010.
64. Bianco, A.; Mantero, M.; Oggioni, L.; Pariani, G.; Bertarelli, C. Advances in photochromic computer-generated holograms. *Proc. SPIE* **2019**, *11030*, 110300B.

© 2019 by the authors. Licensee MDPI, Basel, Switzerland. This article is an open access article distributed under the terms and conditions of the Creative Commons Attribution (CC BY) license (http://creativecommons.org/licenses/by/4.0/).

MDPI
St. Alban-Anlage 66
4052 Basel
Switzerland
Tel. +41 61 683 77 34
Fax +41 61 302 89 18
www.mdpi.com

Materials Editorial Office
E-mail: materials@mdpi.com
www.mdpi.com/journal/materials

www.ingramcontent.com/pod-product-compliance
Lightning Source LLC
LaVergne TN
LVHW070646100526
838202LV00013B/896